T0259955

Lecture Notes in Production Engineering

More information about this series at http://www.springer.com/series/10642

Christian Brecher

Editor

Advances in Production Technology

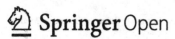

Editor
Christian Brecher
RWTH Aachen
Aachen
Germany

ISSN 2194-0525   ISSN 2194-0533   (electronic)
Lecture Notes in Production Engineering
ISBN 978-3-319-36572-5   ISBN 978-3-319-12304-2   (eBook)
DOI 10.1007/978-3-319-12304-2

Springer Cham Heidelberg New York Dordrecht London

Printed on acid-free paper

Springer International Publishing AG Switzerland is part of Springer Science+Business Media
(www.springer.com)

Preface

CEO of the Cluster of Excellence "Integrative Production Technology for High-Wage Countries"

This edited volume contains the papers presented at the scientific advisory board meeting of the Cluster of Excellence "Integrative Production Technology for High-Wage Countries", held in November 2014 at RWTH Aachen University. The cluster is part of the German Universities Excellence Initiative funded by the German Research Association (DFG) with the aim to contribute solutions to economically, ecologically and socially sustainable production in high-wage countries. To achieve this goal researchers from 27 different institutes in Aachen work on an integrative, discipline-spanning approach combining production engineering, materials science, natural sciences as well as economics and social sciences.

The international scientific advisory board assembles every 2 years. These meetings enable us to reflect and evaluate our research results from an external point of view. Thus, we benefit from comprehensive feedback and new scientific perspectives.

The aim of this volume is to provide an overview of the status of research within the Cluster of Excellence. For details the reader may refer to the numerous further technical publications. The Aachen perspective on integrative production is complemented by papers from members of the international scientific advisory board, all leading researchers in the fields of production, materials science and bordering disciplines.

The structure of the volume mirrors the different projects within the cluster. It includes individualised production, virtual production systems, integrated technologies and self-optimising production systems. These technical topics are framed by an approach to a holistic theory of production and by the consideration of human factors in production technology.

I would like to thank the scientific advisory board for their valuable feedback, especially those members who contributed to the meeting with papers and presentations. Further, I would like to thank the scientists of the cluster for their results and the German Research Foundation (DFG) for the funding and their support.

Aachen, November 2014   Christian Brecher

Contents

Part III Virtual Production Systems

Part IV Integrated Technologies

Part V Self-Optimising Production Systems

Part VI   Human Factors in Production Technology

Chapter 1
Introduction

Christian Brecher and Denis Özdemir

1.1 The Cluster of Excellence "Integrative Production Technology for High-Wage Countries"

Manufacturing is fundamental for the welfare of modern society in terms of its contribution to employment and value added. In the European Union almost 10 % of all enterprises (2.1 million) were classified to manufacturing (Eurostat 2013). With regards to the central role of manufacturing, the European Commission (2012) aims to increase the share of manufacturing from 16 % of GDP (2012) to 20 % by 2020.

Manufacturing companies in high-wage countries are challenged with increasing volatile and global markets, short innovation cycles, cost-pressure and mostly expensive resources. However, these challenges can also open up new business opportunities for companies if they are able to produce customer-specific products at mass production costs and if they can rapidly adapt to the market dynamics while assuring optimised use of resources. Today, the two dichotomies behind those capabilities are not yet resolved: Individual products that match the specific customer demands (scope) generally result in unit costs far above those of mass production (scale). Moreover, the optimisation of resources with sophisticated planning tools and highly automated production systems (planning orientation) mostly leads to less adaptability than achievable with simple and robust value stream oriented process chains (value orientation). Together, the two dichotomies form the polylemma of production (Fig. 1.1).

The research within the Cluster of Excellence aims to achieve sustainable competiveness by resolving the two dichotomies between scale and scope and

C. Brecher · D. Özdemir (✉)
Laboratory for Machine Tools and Production Engineering (WZL)
of RWTH Aachen University, Cluster of Excellence "Integrative Production Technology
for High-Wage Countries", Steinbachstr. 19, 52074 Aachen, Germany
e-mail: d.oezdemir@wzl.rwth-aachen.de

© The Author(s) 2015   1
C. Brecher (ed.), *Advances in Production Technology*,
Lecture Notes in Production Engineering, DOI 10.1007/978-3-319-12304-2_1

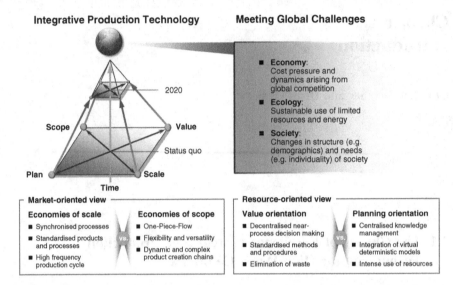

Fig. 1.1 Meeting economic, ecological and social challenges by means of Integrative Production Technology aimed at resolving the polylemma of production (Brecher et al. 2011)

between plan and value orientation (Brecher et al. 2011). Therefore, the cluster incorporates and advances key technologies by combining expertise from different fields of production engineering and materials science aiming to provide technological solutions that increase productivity, adaptability and innovation speed. In addition, sustainable competiveness requires models and methods to understand, predict and control the behaviour of complex, socio-technical production systems. From the perspective of technical sub-systems the complexity can often be reduced to the main functional characteristics and interaction laws that can be described by physical or other formal models. These deterministic models enhance predictability allowing to speed-up the design of products and production processes.

Socio-technical production systems as a whole, however, comprise such a high complexity and so many uncertainties and unknowns that the detailed behaviour cannot be accurately predicted with simulation techniques. Instead cybernetic structures are required that enable a company to adapt quickly and robustly to unforeseen disruptions and volatile boundary conditions. These cybernetic structures start with simple feedback loops on the basis of classical control theory, but also comprise self-optimisation and cybernetic management approaches leading to structural adaption, learning abilities, model-based decisions, artificial intelligence, vertical and horizontal communication and human-machine interaction. The smart factory in the context of "Industrie 4.0" can be seen as a vision in this context (Kagermann et al. 2013). One of the keys for practical implementation of the smart factory will be the understanding and consideration of human factors in production systems (Chap. 14—Brauner and Ziefle).

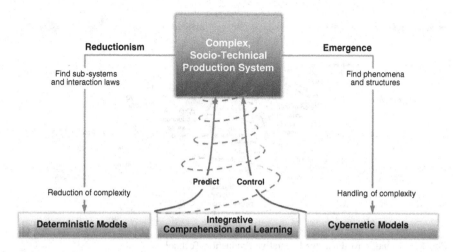

Fig. 1.2 Combining deterministic and cybernetic models

A holistic theory of production to predict and control the behaviour of complex production systems combines deterministic and cybernetic models to enable an integrative comprehension and learning process (Fig. 1.2), e.g. cybernetic approaches that integrate deterministic models or deterministic models that are improved by the feedback within cybernetic structures.

1.2 Scientific Roadmap

To resolve the dichotomies a scientific roadmap with four Integrated Cluster Domains (ICDs) has been defined at the start of cluster in the year 2006 (Fig. 1.3). The research within the domain Individualised Production (ICD A) focusses on the dichotomy between scale and scope. Thus, the main research question is, how small quantities can be manufactured in a significantly more efficient manner by reducing the time and costs for engineering and set-up (Fig. 1.4). A promising approach in this context is Selective Laser Melting (SLM), an additive manufacturing technology that has been significantly advanced within the cluster (Chap. 5—Poprawe et al.). By applying a laser beam selectively to a thin layer of metal powder, products with high-quality material characteristics can be manufactured without tools, moulds or machine-specific manual programming. On this basis individuality can be achieved without additional costs allowing new business models different from those of mass production (Chap. 4—Piller et al.).

While additive manufacturing will be beneficial for certain applications, it will not replace established mould-based technologies. Rather, the aim is to efficiently produce small batches under the constraint that each batch requires a custom mould or die. Time and costs for engineering and set-up can be reduced by applying

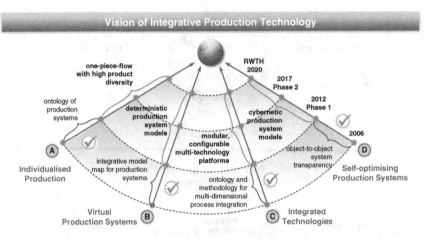

Fig. 1.3 Scientific roadmap for Integrative Production Technology

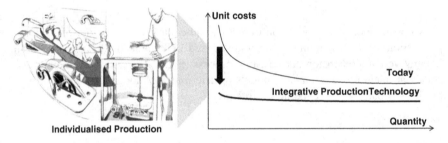

Fig. 1.4 Objective of Individualised Production (Brecher and Wesch-Potente 2014)

simulation-based optimisation methods, instead of being dependent on multiple run-in experiments and expensive modifications (Siegbert et al. 2013). Further, modular parts of moulds or dies can be manufactured by SLM allowing a direct realisation of the results from topology optimisation.

Virtual Production Systems (ICD B) are a prerequisite not only for Individualised Production (ICD A), but also for the design of Integrated Technologies (ICD C) and for the "Intelligence" within Self-optimising Production Systems (ICD D). The research in the field of ICD B addresses the dichotomy between planning and value orientation by developing methods that increase innovation speed and allow a fast adaption to new requirements. Integrative Computational Materials and Production Engineering (ICMPE), for example, provides a platform that can significantly reduce the development time for products with new materials (Chap. 7—Bleck et al.). To fully leverage the potential of simulation-based approaches, concepts for information aggregation, retrieval, exploration and visualisation have been developed in the cluster. Schulz and Al-Khawli demonstrate this approach using the example of laser-based sheet metal cutting, where the dependencies within the high dimensional parameter set are aggregated in a process map (Chap. 6—Schulz and Al-Khawli). On factory level, dependencies are modelled with ontology languages (Büscher et al. 2014) and visualised with Virtual Reality (Pick et al. 2014).

The research within the area of Integrated Technologies (ICD C) aims to combine different materials and processes to shorten value chains and to design products with new characteristics. Integrating different technologies leads to greater flexibility, more potential for individualisation and less resource consumption. Considering production systems, hybrid manufacturing processes enable the processing of high strength materials, e.g. for gas turbines (Lauwers et al. 2014) (Chap. 8—Lauwers et al.). Within the cluster a multi-technology machining centre has been developed in a research partnership with the company CHIRON. The milling machine that is equipped with two workspaces integrates a 6-axis robot and two laser units, one for laser deposition welding and hardening and the other for laser texturing and deburring. Both can be picked up by the robot or by the machine spindle from a magazine (Brecher et al. 2013b). Research questions comprise the precision under thermal influences, control integration, CAM programming, safety and economic analysis (Brecher et al. 2013a, 2014). Hybrid sheet metal forming, as another example for integrated technologies, combines stretch-forming and incremental sheet forming allowing variations of the product geometry without the need for a new mould (Chap. 9—Hirt et al.). Multi-technology production systems facilitate the production of multi-technology products that integrate different functionalities and materials in one component. Examples that have been developed within the cluster include microstructured plastics optics, plastic bonded electronic parts and light-weight structural components (Chap. 10—Hopmann et al.) (Fig. 1.5).

Efficient operation of production systems in turbulent environments requires methods that can handle unpredictability and complexity. Self-optimisation

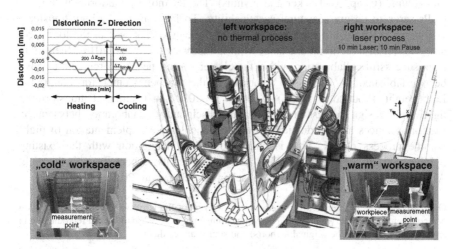

Fig. 1.5 Multi-technology production systems—thermal machine deformation caused by laser-assisted processes (Bois-Reymond and Brecher 2014)

(ICD D) allows dynamic adaptations at different levels of production systems. On the level of production networks the research within the cluster focuses cybernetic production and logistics management (Schmitt et al. 2011). Recent work in this area analyses the human factors in supply chain management (Brauner et al. 2013), an approach that requires the close collaboration of the disciplines engineering, economics and social sciences (Chap. 14—Brauner and Ziefle; Chap. 15—Schmitt et al.). On cell level, human-robot cooperation tasks are considered in a graph-based planner for assembly tasks (Chap. 11—Schlick et al.). To optimise manual assembly tasks with employee-specific support a sound understanding of physiological stress behaviour is required (Graichen and Deml 2014). Graichen et al. from Karlsruhe Institute of Technology (KIT) contribute in this context to the present volume (Chap. 13—Graichen et al.). From a technical perspective self-optimisation has been studied for a wide range of manufacturing processes within the cluster (Chap. 12—Klocke et al.), e.g. injection moulding (Reiter et al. 2014), laser cutting (Thombansen et al. 2014), milling (Auerbach et al. 2013), welding (Reisgen et al. 2014), weaving (Gloy et al. 2013) and assembly (Schmitt et al. 2014). With those practical applications it is demonstrated how self-optimisation helps to achieve cost-efficient production planning and manufacturing.

In addition to the research domains the cluster comprises Cross-Sectional Processes (CSPs) to consolidate the results and to achieve sustainability in terms of scientific, personnel and structural development. For personnel sustainability the CSPs focus activities in the fields of cooperation engineering, innovation management, diversity management and performance measurement (Jooß et al. 2013). For scientific sustainability the CSPs collect and consolidate results and cases from the ICDs for an enhanced theory of production (Chap. 2—Schuh et al.). To complement these results Becker and Nyhuis from the Institute of Production Systems and Logistics (IFA) contribute their framework of a production logistics theory to this volume (Chap. 3—Becker and Nyhuis). The technology platforms within the CSPs serve to ensure structural sustainability. To facilitate technology transfer a web-based platform has been established in the cluster that will in the long term also support bi-directional exchange with industry (Schuh et al. 2013). Stemming from successful collaboration within the cluster several new research centres have been established. A successful example is the Aachen Center for Integrative Lightweight Production (AZL)—funded in 2012—that aims at transforming lightweight design in mass production. Interdisciplinary collaboration between the material sciences and production technology enables the implementation of high-volume process chains. This is carried out in collaboration with the existing lightweight activities of the RWTH Aachen University, especially with the eight AZL partner institutes from RWTH Aachen (Brecher et al. 2013c).

Acknowledgment The authors would like to thank the German Research Foundation DFG for the kind support within the Cluster of Excellence "Integrative Production Technology for High-Wage Countries.

References

Auerbach T, Rekers S, Veselovac D, Klocke F (2013) Determination of characteristic values for milling operations using an automated test and evaluation system. Advanced Manufacturing Engineering and Technologies NEWTECH 2013 Stockholm, Sweden 27–30 October 2013

Bois-Reymond F, Brecher C (2014) Integration of high-performance laser machining units in machine tools: effects and limitations. Proc. Cluster Conferences "Integrative Production Technology for High-wage Countries":93–106

Brauner P, Runge S, Groten M, Schuh G, Ziefle M (2013) Human factors in supply chain management. In: Human Interface and the Management of Information. Springer, pp 423–432

Brecher C, Breitbach T, Do-Khac D, Bäumler S, Lohse W (2013a) Efficient utilization of production resources in the use phase of multi-technology machine tools. Prod. Eng. Res. Devel. 7(4):443-452. doi: 10.1007/s11740-013-0455-5

Brecher C, Breitbach T, Du Bois-Reymond F (2013b) Qualifying Laser-integrated Machine Tools with Multiple Workspace for Machining Precision. Proceedings in Manufacturing Systems 8 (3)

Brecher C, Emonts M, Jacob A (2013c) AZL: a unique resource for lightweight production. Reinforced Plastics 57(4):33–35. doi: 10.1016/S0034-3617(13)70125-5

Brecher C, Hirt G, Bäumler S, Lohse W, Bambach M (2014) Effects and Limitations of Technology Integration in Machine Tools. Proc. 3rd International Chemnitz Manufacturing Colloquium (ICMC):413–432

Brecher C, Jeschke S, Schuh G, et al. (2011) The Polylemma of Production. In: Brecher C (ed) Integrative Production Technology for High-Wage Countries. Springer, Berlin, pp 20–22

Brecher C, Wesch-Potente C (eds) (2014) Perspektiven interdisziplinärer Spitzenforschung. Apprimus, Aachen

Büscher C, Voet H, Meisen T, Krunke M, Kreisköther K, Kampker A, Schilberg D, Jeschke S (2014) Improving Factory Planning by Analyzing Process Dependencies. Proceedings of the 47th CIRP Conference on Manufacturing Systems 17(0):38–43. doi: 10.1016/j.procir.2014.01. 142

European Commission (2012) A Stronger European Industry for Growth and Economic Recovery. SWD 299 final. Communication from the Comission to the European Parliament, Brussels

Eurostat (2013) Manufacturing statistics - NACE Rev. 2. http://epp.eurostat.ec.europa.eu/statistics_explained/index.php/Manufacturing_statistics_-_NACE_Rev._2. Accessed 01 Oct 2014

Gloy Y, Büllesfeld R, Islam T, Gries T (2013) Application of a Smith Predictor for Control of Fabric Weight during Weaving. Journal of Mechanical Engineering and Automation 3 (2):29–37

Graichen S, Deml B (2014) Ein Beitrag zur Validierung biomechanischer Menschmodelle. In: Jäger M (ed) Gestaltung der Arbeitswelt der Zukunft. 60. Kongress der Gesellschaft für Arbeitswissenschaft. GfA-Press, Dortmund, pp 369–371

Jooß C, Welter F, Richert A, Jeschke S, Brecher C (2013) A Management approach for interdisciplinary Research networks in a knowledge-based Society - Case study of the cluster of Excellence "Integrative Production technology for high-wage countries". In: Jeschke S, Isenhardt I, Hees F, Henning K (eds) Automation, Communication and Cybernetics in Science and Engineering 2011/2012. Springer, Berlin, pp 375–382

Kagermann H, Wahlster W, Helbig J (eds) (2013) Umsetzungsempfehlungen für das Zukunftsprojekt Industrie 4.0. Abschlussbericht des Arbeitskreises Industrie 4.0. acatech, München

Lauwers B, Klocke F, Klink A, Tekkaya AE, Neugebauer R, Mcintosh D (2014) Hybrid processes in manufacturing. CIRP Annals-Manufacturing Technology 63(2):561–583

Pick S, Gebhardt S, Kreisköther K, Reinhard R, Voet H, Büscher C, Kuhlen T (2014) Advanced Virtual Reality and Visualization Support for Factory Layout Planning. Proc. of „Entwerfen Entwickeln Erleben – EEE2014"

Reisgen U, Purrio M, Buchholz G, Willms K (2014) Machine vision system for online weld pool observation of gas metal arc welding processes. Welding in the World:1–5

Reiter M, Stemmler S, Hopmann C, Reßmann A, Abel D (2014) Model Predictive Control of Cavity Pressure in an Injection Moulding Process. Proceedings of the 19th World Congress of the International Federation of Automatic Control (IFAC)

Schmitt R, Brecher C, Corves B, Gries T, Jeschke S, Klocke F, Loosen P, Michaeli W, Müller R, Poprawe R (2011) Self-optimising Production Systems. In: Brecher C (ed) Integrative Production Technology for High-Wage Countries. Springer, Berlin, pp 697–986

Schmitt R, Janssen M, Bertelsmeier F (2014) Self-optimizing compensation of large component deformations. Metrology for Aerospace (MetroAeroSpace), 2014 IEEE:89–94

Schuh G, Aghassi S, Calero Valdez A (2013) Supporting technology transfer via web-based platforms. Technology Management in the IT-Driven Services (PICMET), 2013 Proceedings:858–866

Siegbert R, Elgeti S, Behr M, Kurth K, Windeck C, Hopmann C (2013) Design Criteria in Numerical Design of Profile Extrusion Dies. KEM Key Engineering Materials 554–557:794–800

Thombansen U, Hermanns T, Molitor T, Pereira M, Schulz W (2014) Measurement of Cut Front Properties in Laser Cutting. Physics Procedia 56:885–891

Part I
Towards a New Theory of Production

Anja Ruth Weber and Denis Özdemir

Production systems will never be fully predictable and boundary conditions become increasingly volatile and complex. This affects especially the outcome of decisions within a production system and requires dynamic system behaviour. Theories help to develop production systems. With regard to engineering and manufacturing sciences there exists specialized knowledge in different areas of production, but the broad technical approach and understanding is not sufficient. Therefore, the long-term goal of the Cluster of Excellence "Integrative Production Technology for High-Wage Countries" is a higher level of integrativity in production technology by developing a common theory of value creation in times of Industrie 4.0 and of highly dynamic system behaviour in production. Thereby, the term 'Industrie 4.0' describes the development of information and communication technology finding its way into production whose potentials the German Government strives to realise by a homonym project. Industrie 4.0 presents an enormous challenge such as big data, data processing, data security and human–machine collaboration. A holistic theory of production shall help to design and operate production systems in such an environment, and that in consideration of economic, ecologic and social aspects.

For better modelling real-world correlations the theory should on the one hand describe deterministic processes such as physical and mathematical model chains. By focusing such predictable relationships a reduction of complexity can be achieved. On the other hand the theory needs to adopt a cybernetic perspective, in order to include non-predictable processes. In a real-world situation and within a complex system ambient conditions are changing continuously and decisions often need to be taken, despite the lack of information. This approach enables controlling the complexity, although the situation is not fully understood. Linking the deterministic and cybernetic approach aims at permanently finding the optimised operating point not only regarding economic but also technical aspects.

Existing production theories, however, are rather economic science-oriented. This is related to the historical background. The first attempts of developing a production theory date back to the eighteenth century when TURGOT varied the labour input due to agricultural problems (Fandel 2005). Thus, models from manufacturing engineering need to be complemented. The main challenge lies in the incorporation of the diverse engineering sub-disciplines to a theoretical, descriptive model. Its expansion with economic inputs and outputs then leads to a new theory of production by combining economic and technical aspects.

Reference

Fandel G (2005) Produktion I. Produktions- und Kostentheorie, 6. Aufl. Springer, Berlin

Chapter 2
Hypotheses for a Theory of Production in the Context of Industrie 4.0

Günther Schuh, Christina Reuter, Annika Hauptvogel and Christian Dölle

Abstract Significant increase in productivity of production systems has been an effect of all past industrial revolutions. In contrast to those industrial revolutions, which were driven by the production industry itself, Industrie 4.0 is pushed forward by an enormous change within the current society due to the invention and frequent usage of social networks in combination with smart devices. This new social behaviour and interaction now makes its presence felt in the industrial sector as companies use the interconnectivity in order to connect production systems and enhance collaboration. As employees bring their own smart devices to work the interconnectivity is brought into the companies as well and Industrie 4.0 is pushed into the companies rather than initiated by the companies themselves. On top of productivity improvement within production the fourth industrial revolution opens up new potentials in indirect departments such as engineering. This focus differentiates Industrie 4.0 from the first three industrial revolutions, which mainly focused on productivity increase by optimising the production process. Within the Cluster of Excellence "Integrative Production Technology for High-Wage Countries" of the RWTH Aachen University four mechanisms were developed which describe Industrie 4.0. The mechanisms "revolutionary product lifecycles", "virtual engineering of complete value chains", "better performing than engineered" and "revolutionary short value chains" can be achieved within an Industrie 4.0-environment. This environment is based on the four enablers "IT-Globalisation", "single source of truth", "automation" and "cooperation" and enhances collaboration productivity. Therefore the present paper examines and introduces hypotheses for a production theory in the context of Industrie 4.0. For each mechanism two hypotheses are presented which explain how the respective target state can be achieved. The transmission of these mechanisms into producing companies leads to an Industrie 4.0 capable environment strengthening competitiveness due to

G. Schuh · C. Reuter · A. Hauptvogel · C. Dölle (✉)
Laboratory for Machine Tools and Production Engineering (WZL), RWTH Aachen University, Steinbachstr. 19, 52074 Aachen, Germany
e-mail: c.doelle@wzl.rwth-aachen.de

C. Brecher (ed.), *Advances in Production Technology*,
Lecture Notes in Production Engineering, DOI 10.1007/978-3-319-12304-2_2

11

increased collaboration productivity within the direct and especially indirect departments. The specified hypotheses were developed within the framework of the Cluster of Excellence "Integrative Production Technology for High-Wage Countries" of the RWTH Aachen University.

2.1 Introduction

This paper continues the work described in "Collaboration Mechanisms to increase Productivity in the Context of Industrie 4.0" (Schuh et al. 2014a). Therefore the present paper proceeds by giving a short introduction regarding Industrie 4.0-enablers. Each mechanisms presented in Schuh et al. (2014a) is then briefly described before two hypotheses for each mechanism are introduced.

The effect of past industrial revolutions has always been a significant increase in productivity (Schuh et al. 2013a). The increase in productivity started with the first industrial revolution due to the introduction of the steam engine and continued with the Taylorism and the automation as well as computerising (Schuh et al. 2013a, 2014b). Thus automation and computerising already increased productivity within the indirect departments the first three industrial revolutions mainly took place on a shop-floor level. Industrie 4.0 continues to shift the productivity increase even more, as especially indirect departments such as engineering are enhanced due to the Industrie 4.0-enablers and further support of software (Russwurm 2013). Therefore this industrial revolution supports decision making, simulation and engineering performance by aid of collaboration. The mentioned performance increase is represented by four mechanisms of increased productivity, which are supported by the Industrie 4.0-enablers (Schuh et al. 2014a).

This paper reflects the mechanisms of productivity increase and introduces hypotheses on how these target states are to be achieved within an Industrie 4.0-environment.

2.2 Collaboration Productivity Due to Industrie 4.0-Enablers

Within the literature the industrial change due to the fourth industrial revolution addresses diverse aspects of Industrie 4.0 and therefore differs widely in its interpretation (Wahlster 2013; Brettel et al. 2014a; Imtiaz and Jasperneite 2013). Still most of the authors agree with the high potential of productivity increase which accompanies the current transformation process. As stated earlier Industrie 4.0 is not initiated on a shop-floor level and therefore companies have to take measures in their own hands to introduce Industrie 4.0-enablers into their companies to profit from the current change in society and technology (Kagermann et al. 2013).

Fig. 2.1 Enabler of collaboration productivity (Schuh et al. 2014a)

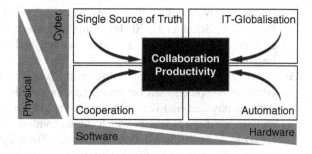

These measures can be categorised by different preconditions which are to be created within a production system. The categorisation is conducted by aid of two dimensions. The first dimension describes whether a precondition is physical or cyber, whereas the second dimension allocates the precondition to hard- or software components (Schuh et al. 2014a). By making up a matrix of the named dimensions four main preconditions can be identified which are shown in Fig. 2.1 and represent the enablers for Industrie 4.0: IT-Globalisation, single source of truth, automation and cooperation.

In order to benefit from the fourth industrial revolution, the presented enablers for collaboration productivity and thus for Industrie 4.0 have to be focused and put into use as a technological and organisational foundation. Against the background of the dimensions for the enablers of Industrie 4.0 collaboration is seen as the interworking of human and human, machine and human and machine and production system (Schuh et al. 2013a, 2014c).

In the following the four enablers for Industrie 4.0 are described as they make up the basis for the productivity increase in an Industrie 4.0-environment as well as the mechanisms and therefore the hypotheses which represent the main focus of this paper.

(1) *IT-Globalisation* The intersection of cyber and hardware concentrates on the IT-Globalisation. Computers present potentials and advantages for economic growth in comparison to the investment costs (Brynjolfsson and Hitt 2000; Schuh et al. 2014a). In the near future the speed of computers will increase even more and therefore becomes less expensive just as storage capacity (Hilbert and López 2011). This will especially enhance producing companies to store massive information in a central cloud which can be accessed from all over the world due to increased speed (Schuh et al. 2014a). On top the increased speed will allow faster extensive simulations of different aspects of a company as well as the processing of huge amounts of data, which are already collected by companies, but cannot be used adequately.

(2) *Single source of truth* To receive viable simulations and information it is inevitable for a company to embed all product lifecycle data along the value chain within a single database (Schuh et al. 2011). Consistent information within this "single source of truth" has to be maintained in terms of product lifecycle management (PLM) to make all changes to product and production

visible and avoid ambiguity (Gecevska et al. 2012; Eigner and Fehrenz 2011; Bose 2006; Schuh et al. 2014a). "Single source of truth" is enhanced by the enabler IT-Globalisation, as cloud storage and access is supported and improved.

(3) *Automation* Further enabler for Industrie 4.0 are cyber-physical systems which combine computers, sensors and actuators and therefore link up the virtual with the physical environment (Lin and Panahi 2010). This leads to automated and decentralised processes which can be combined to collaboration networks (Frazzon et al. 2013; Schuh et al. 2014a). These cyber-physical systems are able to adapt to dynamic requirements and therefore are self-optimising (Wagels and Schmitt 2012). Next to the improvement of machine collaboration this enabler empowers the embedment of skilled workers in such a machine system and enables even more flexible production processes (Schuh et al. 2014a).

(4) *Cooperation* The fourth and therefore last enabler for Industrie 4.0 is called cooperation and aims at the connection of all technologies and activities. Cooperation is already used in development projects, as for example a major NASA supplier named Thiokol achieved a reduction of development lead time by 50 % due to efficient sharing and exchange of engineering data within a network of engineers (Lu et al. 2007). Networks help to improve cooperation by communicating targets and empowering decision maker's in decentralised systems (Kagermann et al. 2013; Schuh et al. 2014a).

The presented enabler depend on each other and also enhance one another as for example simulations using big data is only possible by adequate storage capacities and computing speed. Also automation and collaboration of machines and humans is not possible without the necessary cooperation. In conclusion Industrie 4.0 can only be achieved by developing and applying all four enablers simultaneously (Schuh et al. 2014a).

2.3 Mechanisms and Target States Due to Increased Productivity

The proposed enabler for an Industrie 4.0-environment help to increase the (collaboration) productivity significantly. This significant increase is represented by the four mechanisms "Revolutionary product lifecycles", "Virtual engineering of complete value chains", "Revolutionary short value chains" and "Better performing than engineered" (Schuh et al. 2014a). In the following for each one of the mechanisms hypotheses are presented which propose how the target state, represented by the mechanism, is to be achieved and how Industrie 4.0-enabler help to achieve them.

2.3.1 Revolutionary Product Lifecycles

In today's business environment producing companies face the challenges of shorter lifecycles and micro segmentation of markets (Schuh 2007). Therefore it is essential for such companies to maintain and maybe even extend their development and innovation productivity (Schuh et al. 2013b). One performance indicator for a company's innovation productivity is the time to market. The faster a company is able to introduce new products to the market the shorter the development process has to be. This compression of the development process is made possible within an Industrie 4.0-environment (Schuh et al. 2014a). By aid of integrated technologies and rapid prototyping companies are able to produce testable prototypes which supply viable information of the products potentials as customer feedback can be implemented immediately. Due to the new technologies the costs of an iteration and the resulting changes are not as cost intensive as before and therefore lead to a new development process in terms of time and profit which is shown in Fig. 2.2 (Rink and Swan 1979).

The adjustment of the product development process in terms of profit and time can be achieved by adapting the following hypotheses:

(1) *"Trust based and iterative processes are more productive and more efficient than deterministically planned processes"*

Trust based and iterative processes lead to an increase in productivity as developers are afforded time and space to invent, albeit within set boundaries, and therefore generate more innovations than within a deterministically planned process (Paasivaara et al. 2008; Schuh et al. 2014a). As the new development process is based on a SCRUM-like approach, deterministic planning becomes less important as iterations are permitted and also promoted (Schwaber and Beedle 2002; Schuh et al. 2014a). Thus planning a whole development process would take up a huge amount of time considering all possible solutions within the design space. Unlike nowadays the iterations and adaptations due to field tests are not as cost intensive as new technologies such as selective laser melting and rapid prototyping offer "complexity for free" and are able to generate new prototypes in significant less time and with less recourses.

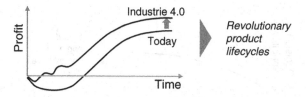

Fig. 2.2 Revolutionary product lifecycles (Schuh et al. 2014a)

(2) *"The speed of a planning process is more important than the quality of the planning process itself"*

The second hypothesis mainly aims at the planning process within product development projects. Nowadays projects are accurately planned, which takes up a great amount of time and also causes analogous costs within a state where a lot of uncertainty is common due to unknown risks within the development process. Therefore the current process is also based on the assumption that adaptations and alteration to the project are to be prevented (Brettel et al. 2014b). However, the development process within the Industrie 4.0-environment supports iterations and therefore alterations. Thus it is more important to quickly generate a plan in order to start the next development step than to accurately predict the outcome of this development step (Gilbreth 1909; Mees 2013). Furthermore the new integrated production technologies allow adaptations which might be necessary due to unforeseen events.

2.3.2 Virtual Engineering of Complete Value Chains

Software tools such as OptiWo are able to virtualise global production networks and help to optimise the production setup (Schuh et al. 2013c). By aid of such tools companies now have the opportunity to simulate their whole production network. This virtualisation and simulation can reveal possible capacity problems as well as problems within the general workflow (Schuh et al. 2014a). By simulating the value chain in a short amount of time one is able to counteract possible problems before they arise, which enhances the decision capability. Furthermore the virtualisation of the value chain supports product development, as the effects of measures taken in the early stages of a product's lifecycle can be simulated and evaluated. The prediction of possible problems due to faults within product development contains a high cost potential as the error correction costs increase exponentially over time (Pfeifer 2013). Therefore the virtualisation enhances the iterative development and consequently also the radically short development processes as virtual try-out is supported (Takahashi 2011). To get a valuable decision capability based on simulations it is necessary to execute an adequate number of simulations (Fig. 2.3).

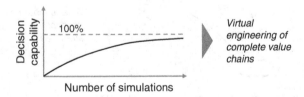

Fig. 2.3 Virtual engineering of complete value chains (Schuh et al. 2014a)

(1) *"The quality of planning decisions is enhanced by a fast development of the complete virtual value chain"*

In order to get an even better decision making capability it is very important to gain information as fast and early as possible. Even in an Industrie 4.0-environment with high speed computers simulation takes time and different situations have to be generated. Furthermore the rule of ten states that costs for error correction increase exponentially (Pfeifer 1996). Therefore the fast implementation of a virtual value chain helps to start simulating as early as possible in order to detect possible errors which in a next step can be addressed by adequate measures. This results into better planning decisions and results due to preventive measures.

(2) *"Increasing the number of different simulation scenarios improves decision making due to better understanding and examination of assumptions"*

Following the law of large numbers in which the accuracy of the relative probability is increased by an infinite number of attempts, the amount of simulations for a specific situation within the value chain effects the capability to make right decisions. The logical implication being, that with an increasing number of simulation scenarios the actual outcome of a given set up of for example a manufacturing process and its ambient conditions will be detected and therefore the right measures can be taken. In analogy to the law of great numbers of Bernoulli where increasing the number of experiments leads to a higher accuracy (Albers and Yanik 2007; Schuh et al. 2014a) this hypothesis states, that the possibility of simulating the future case increases adequately and therefore the outcome of the future scenario is known due to the simulation and therefore can be taken into account for the decision. In combination with the Industrie 4.0-enabler "Speed" the basis of a decision can be improved even more as a computer is able to rapidly combine the results of the simulation.

2.3.3 Revolutionary Short Value Chains

As described before, companies have to offer more and more individualised products in order to meet the customer requirements. As an example of the automobile industry the Ford Fusion is offered in over 15 billion different configurations (Schleich et al. 2007). This trend complicates the division of labour introduced by Taylorism in terms of production and assembly lines, as machines in general are only able to fullfil one specific task. Therefore the complexity of the whole production system is increased. In order to allow even more individualised products the integration of production steps and thus the integration of functions within production systems is inevitable. This leads to a reversion of Taylorism implemented during the second industrial revolution. Instead of the division of labour by means of a conveyor belt production cells are to be established, allowing an employee to

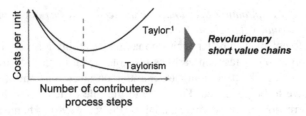

Fig. 2.4 Revolutionary short value chains (Schuh et al. 2014a)

take over autonomous responsibility and give this specific employee decision capability (Schuh et al. 2014a).

Within a production process for highly customised products there is an optimal number of contributors or process steps in one production cell which have to collaborate in order to achieve minimal costs for the produced product (Fig. 2.4).

(1) *"Shortening the process chain by aid of integrated technologies increases productivity"*

Especially within machinery and plant engineering products are produced within a job shop production process. The results of several analyses of the Laboratory for Machine Tools and Production Engineering (WZL), especially in companies with individual and small series production, demonstrated that by passing on the product to the next manufacturing and production step a lot of time elapses due to set up time and downtimes of the machines. As the process chain becomes longer the respective setup and downtimes become longer as well. Long process chains are often caused by the inability to process a unit within one production cell. By integrating different technologies into one machine within an Industrie 4.0-environment the possibility arises to process one specific product within a single or at least a few production cells. Thereby the value chain could be shortened in order to reach a minimum costs per unit by eliminating set up and machine downtime.

(2) *"Continuous process responsibility increases the productivity of the processes"*

As stated before, many companies face the challenge of more and more individualised products. Within Industrie 4.0 it is conceivable that customisation will be taken even further (Brecher et al. 2010; acatech 2011) and companies will not only have to produce customised products of the same kind such as cars, but will have to manufacture totally different products. In this case it is hardly possible to divide the production and manufacturing process into smaller parts in terms of Taylorism. In order to still be able to increase productivity one option is the continuous responsibility of one employee for the whole value creation process of one specific unit of a product. This approach has advantages especially if enhanced by Industrie 4.0. First of all in combination with integrated technologies and processes the continuous responsibility will lower inefficiencies in terms of set up times on the side of the employee as handovers are reduced and the new employee doesn't have to adapt to the specialties of the customised product. As mistakes mostly occur during handovers a continuous responsibility also prevents these mistakes (Prefi 2003). Secondly the responsibility for a whole value creation process gives the employee pride in the product he produces as he sees the development of the product. It was shown, that it is important for an employee to see the results of his work, that the results were impacted by his skills, that they solved difficult problems and that they felt they were trusted (Nakazawa 1993). It is easy to imagine, that the above mentioned feelings are hard to achieve, if the production process is divided into many small steps due to Taylorism. Therefore a continuous process responsibility can help to increase motivation and therefore productivity. This kind of attachment and motivation to increase productivity is already used within the engine manufacturing process at Mercedes-AMG where one single engine is handcrafted and even signed by one single engineer (Höltkemeier and Zwettler 2014).

2.3.4 Better Performing Than Engineered

The mechanism of "Better performing than engineered" aims at the self-optimising capabilities of production systems which are already theoretically possible (Schuh et al. 2013d). With the ongoing advancement of self-optimising production systems machines should be able to reach a productivity level which exceeds the previously determined maximum due to cybernetic effects (Schuh et al. 2014a). These effects would involve structural changes to a system as a response to varying conditions appealing to the production system. An example for such a self optimisation would be a productivity of 15,000 units whereas the estimated maximum before self optimisation was 10,000 units. This kind of self optimisation would have a huge impact on the flexibility and reactivity of a production system and therefore contribute significantly to its productivity. The described self-optimising effect is shown in Fig. 2.5.

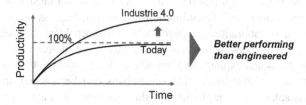

Fig. 2.5 Better performing than engineered (Schuh et al. 2014a)

(1) *"When a self-optimising system reaches its process performance limits the self-optimisation constitutes a process pattern change"*
    In general systems of all kinds are optimised within the systems current state in order to reach an optimal performance level. Usually this level is approached by a decreasing speed. Whenever the optimal performance level is reached no further optimisation is possible. The only way to improve performance beyond this theoretical border is a change within the system itself or within the process pattern. An example for this kind of optimisation is represented by the Fossbury Flop whereas the jumping height could not be improved by the old jumping technique the Fossbury Flop enabled athletes to reach new records. For a production system this pattern change describes the dynamic adaption of the target system. The production system does not only try to reach an exogenous given target but adjusts this target based on internal decisions (Schmitt and Beaujean 2007). Within Industrie 4.0 self-optimising systems therefore should be able to acknowledge performance boarders and change process patterns in order to surpass them.

(2) *"Self-optimisation requires an over determined sensor actuator system"*
    The term "determined" states the described system is fixed within its pre determined patterns, as no degrees of freedom are available to the system to adapt its patterns. For an over-determined system however, there is a possibility to change patterns. For example within a pattern change one degree of freedom can be taken away in exchange for another degree of freedom. Thus a system can adapt to changing requirements. This type of learning and adaption requires a cognitive system, which contains sensors and actuators (Zaeh et al. 2010). Nowadays the change within patterns is usually supported by a human worker (Schmitt et al. 2007), who then expands the sensor actuator system of the production system. To replace the human intervention it is therefore necessary to provide the self-optimising systems with an over-determined sensor actuator system.

2.4 Conclusion

This paper pursues the vision that one core characteristic of Industrie 4.0 is a raise in collaboration productivity. Accordingly, four main enablers as preconditions for Industrie 4.0 and collaboration are introduced. These enablers can help to reach mechanisms or target states, which represent a significant increase in productivity. The paper introduces and explains two hypotheses for each of the four mechanisms, which indicate how the Industrie 4.0-mechanisms can be reached and how the Industrie 4.0-enablers help implementing the mechanisms. Future research will focus on the empirical validation of the depicted hypotheses and mechanisms in order to strengthen or adapt the pursued vision.

Acknowledgments The authors would like to thank the German Research Foundation DFG for the kind support within the Cluster of Excellence "Integrative Production Technology for High-Wage Countries". The authors would further like to thank Anja Ruth Weber and Jan-Philipp Prote for their collaboration in the project.

References

acatech (2011) Cyber-Physical Systems, Driving force for innovation in mobility, health, energy and production (acatech Position paper).

Albers R, Yanik M (2007) Binomialverteilung, In Skript zur Vorlesung „Stochastik", Universität Bremen, http://www.math.uni-bremen.de/didaktik/ma/ralbers/Veranstaltungen/Stochastik12/.

Bose R (2006) Understanding management data systems for enterprise performance management, In Industrial Management & Data Systems 106 (1), pp. 43–59.

Brecher C, Jeschke J, Schuh G, Aghassi S, Arnoscht J, Bauhoff F, Fuchs S, Jooß C, Karmann W. O, Kozielski S, Orilski S, Richert A, Roderburg A, Schiffer M, Schubert J, Stiller S, Tönissen S, Welter F (2010) Individualised Production, In Integrative Production Technology for High-Wage Countries, Berlin: Springer, pp. 77–239.

Brettel M et al. (2014b) Increasing the Problem Solving Speed Through Effectual Decision Making, In AOM 2014: The 74th Annual Meeting of the Academy of Management - August 1–5, 2014, Philadelphia, PA.

Brettel M, Friederichsen N, Keller M, Rosenberg M (2014a) How Virtualization, Decentralization and Network Building Change the Manufacturing Landscape: An Industry 4.0 Perspective, In International Journal of Mechanical, Industrial Science and Engineering 8 (1), pp. 37–44.

Brynjolfsson E, Hitt L M (2000) Beyond Computation: Information Technology, Organizational Transformation and Business Performance, In The Journal of Economic Perspectives 14 (4), pp. 23–48.

Eigner M, Fehrenz A (2011) Managing the Product Configuration throughout the Lifecycle, In 8th International Conference on Product Lifecycle Management, Seoul, pp. 396–405.

Frazzon E, Hartmann J, Makuschewitz T, Scholz-Reite B (2013) Towards Socio-Cyber-Physical Systems in Production Networks, In 46th CIRP Conference on Manufacturing Systems 2013 7 (0), pp. 49–54.

Gecevska V, Veza I, Cus F, Anisic Z, Stefanic N (2012) Lean PLM – Information Technology Strategy for Innovative and Sustainable Business Environment, In International Journal of Industrial Engineering and Management 3 (1), pp. 15–23.

Gilbreth F (1909) The bricklaying System, The MC Clark publishing co, New York.

Hilbert M, López P (2011) The World's Technologica Capacity to Store, Communicate, and Compute Information, In Science 1 April 2011 (332), pp. 60–65.

Höltkemeier U, Zwettler M (21.02.2014) High-Speed-Junkie für einen Tag, In Konstruktionspraxis February 2nd 2014, http://www.konstruktionspraxis.vogel.de/themen/antriebstechnik/motoren/articles/435236/index2.html (accessed on 10 September 2014).

Imtiaz J, Jasperneite J (2013) Scalability of OPC-UA Down to the Chip Level Enables "Internet of Things", In 11th IEEE International Conference on Industrial Informatics, Bochum, pp. 500–505.

Kagermann H, Wahlster W, Helbig J (2013) Recommendations for implementing the strategic initiative Industrie 4.0, Acatech, pp. 13–78.

Lin K-J, Panahi M (2010) A Real-Time Service-Oriented Framework to Support Sustainable Cyber-Physical-Systems, In IEEE 8th International Conference on Industrial Informatics 2010, Osaka, pp. 15–21.

Lu S C-Y, ElMaraghy W, Schuh G, Wilhelm R (2007) A Scientific Foundation of Collaborative Enginereering, In CIRP Annals - Manufacturing Technology 56 (2), pp. 605–634.

Mees B (2013) Mind, Method and Motion, In The Oxford handbook of management theorists, pp. 32–48.

Nakazawa H (1993) Alternative Human Role in Manufacturing, In AI & SoC (1993) 7, pp. 151–156.

Paasivaara M, Durasiewicz S, Lassenius C (2008) Distributed Agile Development: Using Scrum in a Large Project, In 2008 IEEE International Conference on Global Software Engineering, Bangalore, pp. 87–95.

Pfeifer T (1996) Qualitätsmanagement - Strategien, Methoden, Techniken, München: Hanser.

Prefi T (2003) Qualitätsorientierte Unternehmensführung, P3 - Ingenieurges. für Management und Organisation.

Rink D R, Swan J E (1979) Product life cycle research: A literature review, In Journal of Business Research 7 (3), pp. 219–242.

Russwurm S (2013) Software: Die Zukunft der Industrie, In Industrie 4.0 – Beherrschung der industriellen Komplexität mit SysLM, pp. 21–36.

Schleich H, Schaffer J, Scavarda L F (2007) Managing Complexity in Automotive Production, In 19th International Conference on Production Research 2007, Valparaiso, Chile.

Schmitt R et al (2007) Self-optimising Production Systems, In Integrative Production Technology for High-Wage Countries, Berlin: Springer, pp. 697–971.

Schmitt R, Beaujean P (2007) Selbstoptimierende Produktionssysteme - Eine neue Dimension von Flexibilität, Transparenz und Qualität, In ZWF Jahrg. 102 (2007), pp. 520–524.

Schuh G (2007) Excellence in production, Apprimus Wissenschaftsverlag, Aachen.

Schuh G, Dölle C, Riesener M, Rudolf S (2014b) Lean Innovation durch globales Komplexitätsmanagement, In Industrie 4.0: Aachen Perspektiven, Aachener Werkzeugmaschinenkolloquium 2014, pp. 145–170.

Schuh G, Potente T, Fuchs S, Thomas C, Schmitz S, Hausberg C, Hauptvogel A, Brambring F (2013d) Self-Optimising Decision-Making in Production Control, In Robust Manufacturing Control, Berlin: Springer, pp. 443–454.

Schuh G, Potente T, Kupke D, Varandani R (2013c) Innovative Approaches for Global Production Networks, In Robust Manufacturing Control, Berlin: Springer, pp. 385–397.

Schuh G, Potente T, Varandani R, Hausberg C, Fränken B (2014c) Collaboration Moves Productivity To The Next Level, To be published in 47th CIRP Conference on Manufacturing Systems 2014.

Schuh G, Rudolf S, Arnoscht J (2013b) Product-Oriented Modular Platform Design, In Coma '13 International Conference on Competitive Manufacturing Proceedings.

Schuh G, Stich V, Brosze T, Fuchs S, Pulz C, Quick J, Schürmeyer M, Bauhoff F (2011) High resolution supply chain management: optimized processes based on self-optimizing control loops and real time data, In Prod. Eng. Res. Devel. 5 (4), pp. 433–442.

Schuh G, Potente T, Wesch-Potente C, Hauptvogel A (2013a) Sustainable increase of overhead productivity due to cyber-physical-systems, In Proceedings of the 11th Global Conference on Sutainable Manufacturing – Innovation Solutions, pp. 332–335.

Schuh G, Potente T, Wesch-Potente C, Weber A R, Prote J-P (2014a) Collaboration Mechanisms to increase Productivity in the Context of Industrie 4.0, In 2nd CIRP Robust Manufacturing Conference (RoMac 2014), pp. 51–56.

Schwaber K, Beedle M (2002) Agile Software Development with Scrum, Pearson Studium, Upper Saddle River, NJ.

Takahashi S (2011) Virtual Tryout Technologies for Preparing Automotive Manufacturing, In Transactions of JWRI 2012, Osaka.

Wagels C, Schmitt R (2012) Benchmarking of Methods and Instruments for Self-Optimization in Future Production Systems, In 45th CIRP Conference on Manufacturing Systems 2012, pp. 161–166.

Wahlster W (2013) Industry 4.0: The Role of Semantic Product Memories in Cyber-Physical Production Systems, In SemProM: Foundations of Semantic Product Memories for the Internet of Things, Springer, pp. 15–19.

Zaeh M F, Reinhart G, Ostgathe M, Geiger F, Lau C (2010) A holistic approach for the cognitive control of production systems, In Advanced Engineering Informatics 24 (2010), pp. 300–307.

Chapter 3
The Production Logistic Theory as an Integral Part of a Theory of Production Technology

Julian Becker and Peter Nyhuis

> *He who loves practice without theory is like the sailor who boards ship without a rudder and compass and never knows where he may cast.*
>
> (Leonardo Davinci 1452–1519)

3.1 Motivation

Today's manufacturing companies operate in a turbulent environment. Globalisation, increasing market dynamism and ever shortening product life cycles are just some of the aspects that characterise the steady rise in competitive pressure (Roland Berger Strategy Consultants GmbH 2012; Abele and Reinhart 2011; Sirkin et al. 2004). Moreover, factors such as sustainability and the conservation of natural resources are playing an increasingly important role (BMU 2012; Deutsche Post AG 2010). In order to maintain sustainable production in a turbulent environment, it is necessary to be able to anticipate impending changes and to determine and assess available alternative courses of action. The determination of potential action strategies requires knowledge of how production facilities behave at all levels, including those of production networks, machines and processes. Accordingly, in order to maintain their long-term success, companies must be able to predict, analyse and influence changes and the impacts they have on their production. What this requires is a comprehensive theory by which to achieve a scientific understanding and an integral description of production technology. The development of a production logistic **theory** serves to clearly illustrate both the scientific and the practical benefits of such generally applicable theories. Using the example of production logistic theory, this article seeks to determine the fundamental requirements and challenges that are involved in developing such a theory and

J. Becker · P. Nyhuis (✉)
Hannover Centre for Production Technology (PZH), Institute of Production Systems
and Logistics (IFA), An der Universität 2, 30823 Garbsen, Germany
e-mail: nyhuis@ifa.uni-hannover.de

© The Author(s) 2015
C. Brecher (ed.), *Advances in Production Technology*,
Lecture Notes in Production Engineering, DOI 10.1007/978-3-319-12304-2_3

discusses the necessity of a model-based, holistic description of production. Initial approaches towards a theory of production technology are then indicated and future fields of development identified.

3.2 Theory Development in the Context of Production Technology

The interpretation of the term 'theory' may incorporate different aspects, depending on the scientific-theoretical viewpoint adopted, for which reason a brief definition will first of all be given here. The National Academy of Science in the USA defines a theory as "a well substantiated explanation of some aspect of the natural world that can incorporate facts, laws, inferences, and tested hypotheses" (National Academy of Sciences 1998, p. 5). Hence, theories constitute models of reality, on the basis of which it may be possible to derive recommended courses of action. They are verifiable through observation and remain valid until such time as they are scientifically disproven (Popper 1935).

In the context of production technology, theories play a role in generating knowledge, applying them and disseminating them. They are also used in developing production systems as well as in staff training activities. Interest groups include both scientists and users/operators as well as those in training and further education. An unambiguous definition of input, output and system parameters simplifies the basic understanding of the system. Moreover, the theory supplies explanatory models of system behaviour, and the use of uniform terminology enables communication within and between the various interest groups. When designing systems, the use of theory-based construction rules ensures that the required functions are fulfilled. In ongoing operation, theories support the enhancement of subsystems, as they also do in the coordination of subsystems among each other. This makes it possible to counteract, for instance, an obstruction, uncontrolled vibration or incorrect behaviour in a system. In a seminar situation, theories often serve to illuminate a body of knowledge gained through experience, while supplying explanations of system operation and enabling analysis of technical and logistic systems (Wiendahl et al. 2010; Nyhuis and Wiendahl 2007).

The development of a theory involves passing through several consecutive stages (Fig. 3.1). The first stage of theory formation comprises defining the scope of observation and laying down the content boundaries. It is then possible to derive research questions based on this foundation. After collecting together the necessary materials and available knowledge, models and submodels can be developed and validated by means of experiment and protocol. In turn, the hypotheses thus derived enable laws or partial laws to be defined. The combination of the models and laws thus developed ultimately leads to the formulation of (sub)theories of the scope of observation defined at the outset. The models, laws and hence the overall theory

Fig. 3.1 Development of a theory

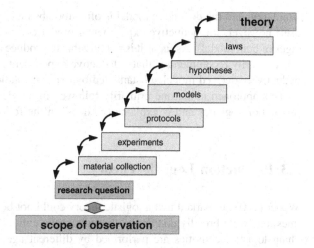

retain their validity until such time as they are scientifically disproven (Nyhuis and Wiendahl 2010).

Model development represents an essential tool in the generation of new knowledge and theories. A number of approaches can be taken in pursuing this end; these are illustrated in Fig. 3.2.

The experimental method involves gaining knowledge by empirical means such as observation and/or experiment. Another approach is to employ the deductive method, in which conclusions are drawn on the basis of purely logical interrelations. The knowledge gained by this means must be empirically verifiable in order for it to be of practical and scientific value. The experimental model frequently only describes a simulation or laboratory experiment for a very specific condition of a

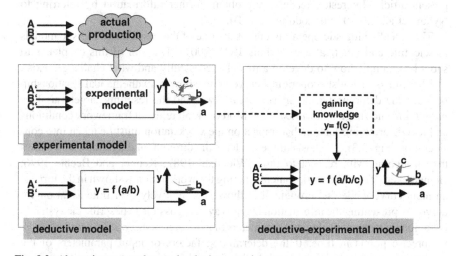

Fig. 3.2 Alternative approaches to developing a model

simulation and the deductive model is often too abstract for the practically oriented user; however, the deductive-experimental model combines the respective advantages of both models. As far as it is not possible to produce a purely deductive model, it is primarily recommended that a deductive-experimental approach be taken and a deductively derived model substantiated by experimentation (Nyhuis et al. 2009).

This approach is the one primarily followed in the development of a theory of production logistics and is explained in detail in the following.

3.3 Production Logistic Theory

Weber (2008) remarked that a logistic theory could not be developed, since logistic questions were broadly distributed throughout the entire value chain. In the supply chain logistics activities are performed by different agents in the various organisational divisions, such as the purchasing, production and distribution departments. A further problem is frequently encountered by virtue of the differences in the target systems in the various divisions. This is precisely why it is urgently necessary to develop a logistic theory that makes it possible to orient the formation of the value chain and the activities of the agents towards a common goal.

The *Institute of Production Systems and Logistics* (*IFA*) has been studying the development of a comprehensive production logistic theory for the internal supply chain for more than 40 years. For this purpose, numerous research questions have been identified and logistic models developed that have been validated in practice and are in broad use. By linking the models, it is possible to conduct a model-based calculation of realisable logistics performance for virtually any configuration within the internal supply chain. Furthermore, basic laws of production logistics and other fundamental laws can be derived (these are however beyond the scope of the present article. Interested readers may obtain further information by referring to Nyhuis et al. (2009) and Lödding (2013)).

The so-called logistic operating curves are one of the best-known logistic models in scientific and practical use (Nyhuis 1991, 2007). The aim of this chapter is to show by example how to develop a model successfully and what challenges exist.

The aim of a logistic operating curve is to show a mathematical relationship between the determining factor *work in process* and the resulting target variables *output rate* and *range of a workstation*. Moreover, all relevant framework conditions and real determining factors that impact on the workstation must be taken into consideration (Fig. 3.3). The basis of the model is the throughput diagram, which shows the throughput with respect to time (Wiendahl 1987; Kettner and Bechte 1976; Heinemeyer 1974). The work content entering the workstation is shown in the form of an input curve while the throughput is shown cumulatively over time as an output curve. By presenting the information in this way, it is possible to describe the system's behaviour in terms of the logistic parameters of *work in process, output rate* and *range* for precise points in time. If the determining factors or input parameters of the workstation, such as capacity levels or lot sizes, change under real conditions, or if

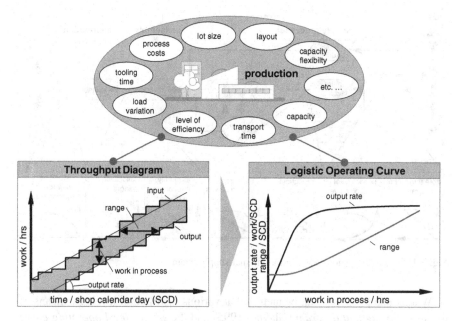

Fig. 3.3 Real factors determining the throughput diagram and logistic operating curves

there are any process disturbances, the effects can be calculated directly and quantitatively. However, the throughput diagram does not—or not fully—describe the cause-effect relationships between the logistic parameters. Hence, one central challenge encountered in deriving logistic operating curves was to determine the cause-effect relationships between the determining factors and the target variables taking into account all determining factors relevant to production. The ideal logistic operating curves were derived in an initial deductive modelling stage. This describes the theoretical limiting values of the logistic key performance indicators with the underlying cause-effect relationships. The modelling of the cause-effect relationships between the performance parameters in real process flows and disturbances was conducted by means of the experimental analysis of simulation results for the purpose of parametric adjustment. This process of deductive-experimental modelling enables the logistic operating curves to adapt easily to changes in framework conditions. Since both the model structure and the input parameters of the model primarily originate from elementary principles, the cause-effect relationships between the logistic target variables can be easily described (Nyhuis et al. 2009).

Figure 3.4 shows examples of further logistic models which, taken together and in the given combination, lead to the formulation of a production logistic theory. The scope of observation comprises the internal supply chain between the procurement and sales markets, which consists of idealised supply, assembly and sales processes. Numerous logistic models have been developed for the respective process elements *store, manufacturing, assembly* and *distribution*. Selected models are briefly presented in the following.

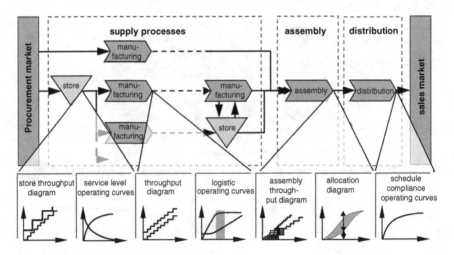

Fig. 3.4 Selected logistic models in the internal supply chain

Warehouse processes can be analysed at various distribution levels by means of the *store throughput diagram* (Gläßner 1995) and the *service level operating curve* (Lutz 2002). These models support to indicate service levels and existing potentials. Moreover, article-specific safety stock can be dimensioned.

As outlined above, manufacturing processes can be described qualitatively and for precise points in time by means of the *throughput diagram*. Building on this, the *logistic operating curves* show the functional dependencies between the logistic target variables *output rate, throughput time, inter-operation time* and *range*. The progression of these target variables is shown as a function of the work in process. This enables a controlling process for the logistic analysis and enhancement of existing production processes. The central challenge encountered in assembly processes lies in the logistic coordination of upstream processes, which can be analysed by means of the *assembly throughput diagram* (Münzberg et al. 2012; Schmidt 2011). The *allocation diagram* (Nyhuis et al. 2013; Beck 2013; Nickel 2008) supplies quantitative statements regarding the consequence of delayed supply from the upstream processes. Moreover, it enables potential to be identified in terms of inventory costs and delivery reliability with regard to assembly processes. Delivery reliability and schedule compliance are particularly important purchasing criteria with respect to the sales market. The *schedule compliance operating curves* (Schmidt et al. 2013) enable an analysis of the scheduling situation of external suppliers or customer-supplier relations within the company, and describe the interrelations between schedule adherence, safety time and stock.

It is now apparent that the logistic models have led to a consistent understanding of the system that constitutes the internal supply chain. This makes it possible to describe, predict and influence logistic system behaviour with respect to logistic parameters such as work in process or lateness. On the basis of these known cause-effect relationships, it is possible to implement a theoretically grounded means of supporting decision-making processes in companies. Models such as the determination of lot-size, safety stock or scheduling of production orders have proven to be of immense practical and sustainable benefit in industrial practice.

It must be borne in mind that the models presented here should be regarded as partial models. Further work is currently being conducted to develop link-variables between the models, by which it will be possible to connect the partial models to form a complete production logistic theory.

Provided the relevant research gaps in the scope of observation can be detected and closed in the near future, it will be possible to gradually expand it. It is therefore conceivable that the theory of production logistics might be extended to incorporate the external supply chain or other target fields such as ecology.

As already stated in the foregoing, the IFA has been involved in researching into a theory of production logistics for the past 40 years. A conspicuous aspect of this is that the development intervals that lead to the formulation of new models are becoming increasingly shorter. While the interval that lay between publications relating to the throughput diagram (Heinemeyer 1974) and those referring to the logistic operating curves (Nyhuis 1991) was as long as 17 years, just under three years separated the development of the assembly throughput diagram (Schmidt 2011) from the analytical description of the allocation diagram (Beck 2013). One of the main reasons for this is the increasing degree of understanding of theory and model development that prevails at the IFA, the effect of which is to considerably accelerate the development process. It is also apparent that the process of theory development not only requires experience but also endurance on the part of research institutes and sponsors of research.

The production logistic theory represents an integral part of production logistics. A comprehensive theory of production technology is necessary to allow comprehensive statements to be made and to recommend courses of action above and beyond the subdisciplines. Initial ideas and approaches will be presented in the following.

3.4 Towards a Theory of Production Technology

The overriding objective of production technology is to transform materials into goods that are destined for a sales market. As a technical science, production technology incorporates principles of natural, economic and social sciences as well as humanities (Spur 2006). Owing to the diverse issues involved, the interlinking of theory and practice plays a particularly important role.

For several decades, numerous approaches have been adopted in the field of business administration towards modelling production by means of a general theory. It is hence impossible within this framework to present a comprehensive overview of the state of the art. Reference is therefore made to Dyckhoff (2002) who recently published an excellent overview along with an appeal that production theory should undergo continued development. He defines production as value

Fig. 3.5 Basic model of a
general production theory in
business administration (acc.
to Dyckhoff (2006))

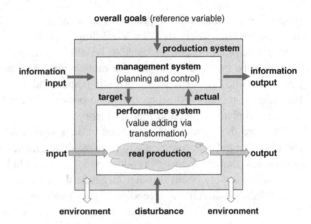

creation that comes about by means of transformation. A transformation is characterised by a qualitative, quantitative, spatial or temporal change of an object or its properties. The basic structure of the general theory of production is illustrated in Fig. 3.5 (Dyckhoff 2006).

A production system consists of two subsystems, the performance system and the management system, which are in a state of interaction with the environment. The performance system comprises the value adding transformation, which is planned, controlled and monitored by the management system. The object under analysis in the general theory of production is the relationship between input and output, which itself can be affected by external disturbance factors. These interrelations are in general formally described by means of production functions. Reference variables are derived from the overall economic goals; these have an effect on the management system. The management system develops targets for the production programme on the basis of incoming information from the market and environment. This is passed on to the performance system, which generates an output from the input variables by means of various transformation processes, in the form of products for the sales market. The actual is reported back to the management system.

A major weakness of the basic model of a general production theory in business administration lies in the functions of production, as these do not explicitly take into account the real value-creation processes. It is not possible to draw direct conclusions about the output of the performance system as a whole from a change brought about to a single production factor, for example the manpower at a workstation. In particular variant rich piece-wise productions of multi-staged products are not practically described by such production functions. Rather, technologically founded transformation steps with varied interaction levels must be taken into consideration. As a rule, resources in the form of raw materials, semi-manufactures and purchased parts are transformed into goods and products in the course of several production and assembly stages. This also requires operative control of production processes, plants and machinery. The scope of production under observation should therefore be expanded to include operative planning and control on the one hand and technological and logistic transformation processes on the other. Furthermore, there are numerous overall goals beside cost-effectiveness to be pursued by businesses in order for them to remain sustainably successful. Logistic performance criteria such as delivery time and delivery reliability must be taken into consideration, as must such factors as flexibility or transformability in the face of short-notice changes to customer requirements. Moreover, ecological and social aspects are becoming increasingly important factors when it comes to fulfilling the overall goal of sustainable production (Nyhuis and Wiendahl 2010; Wiendahl et al. 2010).

The German Academic Society for Production Engineering (WGP) has taken on the task of taking the aforementioned criticisms on board and additionally incorporating a technological production model in the basic model of a general production theory in business administration. Figure 3.6 shows a schematic diagram of this extension.

The technological production model still consists of a management system and a performance system. The overall goal of sustainability has now been incorporated, and alongside economic, ecological and social aspects, functional compliance of the manufactured products has been added. This ensures that production is incorporated as a factor in the fulfilment of customer requirements. The input variables are information, work-force, materials and energy; these are transformed by means of the performance system into elements, parts, part families or sub-products, products

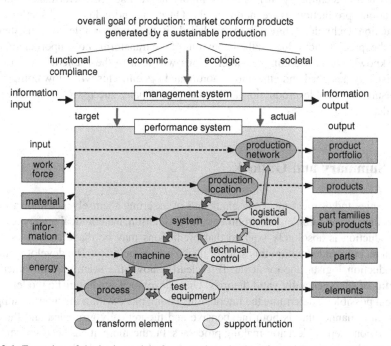

Fig. 3.6 Extension of the basic model of a general production theory in business administration

and overall into a product portfolio. Within the performance system there are transform elements: process, machine, system, production location and production network, to which the support functions of test and measuring equipment, technical control and logistic control have been added.

In order to develop a theory of production technology, it is necessary to determine submodels within and between the transform elements and to derive hypotheses from these. An important first step consists of further subdividing the overall goals. For example, of relevance to functional compliance on the level of elements and parts are the features for denoting the geometric body, its surface, its material properties and its service life. On the product level, the main aspects are functionality, performance and appearance. As for the economic goals, of primary interest are cycle times, tooling times, throughput times and manufacturing costs; delivery times, delivery compliance and the effect on turnover have been added on the upper levels. The ecological goals are characterised by both the material aspect (material utilisation, reutilisation) and the energy aspect; the environmental pollution generated by the production facilities has now been added. Finally, social goals are oriented towards the immediate workplace location and primarily concern their ergonomic and safe design. Of decisive importance at higher levels are such factors as personal communication, and the work content determined by the structural organisation, as well as trust and cooperation (Nyhuis and Wiendahl 2010; Wiendahl et al. 2010).

The WGP model presented here represents a generally valid approach to a theory of production technology. It involves a highly interconnected performance system on various production aggregation levels. Once the individual subdisciplines of production technology have been modelled, it will be possible to consistently describe, predict and influence the behaviour of the transform and support elements. This knowledge of the complex interactions will lead to the implementation of a consistently designed and efficiently coordinated system. This will allow companies to ensure sustainable production and enhance company success even in times of turbulence.

3.5 Summary and Outlook

Companies today must meet the challenge of asserting themselves within a non-deterministic and turbulent environment. For this reason, a theoretical understanding of production is absolutely vital so that predictions may be made and courses of action recommended within this state of increasing complexity. The development of a production logistic theory at the IFA clearly shows the scientific and practical benefits of such generally valid theories. Once developed, such logistic models will make it possible to determine the interactions taking place within the internal supply chain in a manner that is both inexpensive and theoretically grounded and thus to support commercial decision-making processes. Furthermore, it has been shown that a theory development process requires much in the way of both experience and endurance on the part of research institutes and sponsors. Although theoretical approaches relating to production already exist, they are not in themselves sufficient for explaining the complex interactions that take place within production. Accordingly, the scientific and theoretical understanding of production technology as a whole constitutes a major research gap. The WGP approach to a theory of production technology presented here represents an attempt to close this gap by substantiating the scope of observation by means of the technological production model. The aim is to orient the entire performance system with its complex interactions, taking into account the defined overall goals. To this end, further efforts will be required in all subdisciplines to derive research questions from the substantiated overall goals and to develop the respective models. Following this, the individual integral models and theories can be merged to form a theory of production technology.

References

Abele E, Reinhart G (2011) Zukunft der Produktion. Herausforderungen, Forschungsfelder, Chancen. Hanser Verlag, München

Beck S (2013) Modellgestütztes Logistikcontrolling konvergierender Materialflüsse. Berichte aus dem IFA, vol 2013,3. PZH Produktionstechnisches Zentrum, Garbsen

BMU (ed) (2012) GreenTech made in Germany 3.0. Umwelttechnologie-Atlas für Deutschland, Bundesministerium für Umwelt, Naturschutz und Reaktorsicherheit, Berlin

Deutsche Post AG (ed) (2010) Delivering Tomorrow. Towards Sustainable Logistics, Deutsche Post AG, Bonn

Dyckhoff H (2002) Neukonzeption der Produktionstechnologie. ZfB - Zeitschrift für Betriebswirtschaft 73(Heft 3):705–732

Dyckhoff H (2006) Produktionstheorie. Grundzüge industrieller Produktionswirtschaft, 5. überarbeitete Aufl. Springer, Berlin, New York

Gläßner J (1995) Modellgestütztes Controlling der beschaffungslogistischen Prozesskette, Als Ms. gedr., vol 337. VDI Verlag, Düsseldorf

Heinemeyer W (1974) Die Analyse der Fertigungsdurchlaufzeit im Industriebetrieb. Offsetdruck Böttger, Hannover

Kettner H, Bechte W (1976) Neue Wege der Bestandsanalyse im Fertigungsbereich. Methodik, praktische Beispiele, EDV-Programmsystem. Fachbericht des Arbeitsausschusses Fertigungswirtschaft (AFW) der Deutschen Gesellschaft für Betriebswirtschaft (DGfB). Inst. für Fabrikanalgen (IFA) der Techn. Univ, Hannover

Lödding H (2013) Handbook of Manufacturing Control. Fundamentals, description, configuration. Springer, Berlin, Heidelberg

Lutz S (2002) Kennliniengestütztes Lagermanagement. Univ., Diss.–Hannover, 2002, Als Ms. gedr. Reihe 13, Fördertechnik/Logistik, vol 53. VDI-Verl, Düsseldorf

Münzberg B, Schmidt M, Beck S, Nyhuis P (2012) Model based logistic monitoring for supply and assembly processes. Prod. Eng. Res. Devel. 6(4–5):449–458. doi: 10.1007/s11740-012-0403-9

Nickel R (2008) Logistische Modelle für die Montage. Berichte aus dem IFA, 2008, Bd. 02. PZH, Produktionstechn. Zentrum, Garbsen

Nyhuis P (1991) Durchlauforientierte Losgrößenbestimmung, Als Ms. gedr., Nr. 225. VDI Verlag; Düsseldorf

Nyhuis P, Beck S, Schmidt M (2013) Model-based logistic controlling of converging material flows. CIRP Annals - Manufacturing Technology 62(1):431–434. doi: 10.1016/j.cirp.2013.03. 041

Nyhuis P, Wiendahl HP, Rossi R (2009) Fundamentals of production logistics. Theory, tools and applications; with 6 tables. Springer, Berlin

Nyhuis P (2007) Practical Applications of Logistic Operating Curves. CIRP Annals - Manufacturing Technology 56(1):483–486. doi: 10.1016/j.cirp.2007.05.115

Nyhuis P, Wiendahl H P (2010) Ansatz zu einer Theorie der Produktionstechnik. Zeitschrift für wirtschaftlichen Fabrikbetrieb 105(1–2):15–20

Nyhuis P, Wiendahl H P (2007) Ansätze einer Logistiktheorie In: Hausladen I (ed) Management an Puls der Zeit–Strategien, Konzepte und Methoden, 1. Aufl. TCW Transfer-Centrum, München

Popper K (1935) Logik der Forschung. Springer Vienna, Vienna

Roland Berger Strategy Consultants GmbH (2012) Mastering product complexity. http://www. rolandberger.us/media/pdf/Roland_Berger_Mastering-Product-Complexity_20121107.pdf. Accessed 25 Aug 2014

Schmidt M (2011) Modellierung logistischer Prozesse der Montage. Univ., Diss.–Hannover, 2010. Berichte aus dem IFA, vol 2011,1. PZH Produktionstechn. Zentrum, Garbsen

Schmidt M, Bertsch S, Nyhuis P (2013) Schedule compliance operating curves and their application in designing the supply chain of a metal producer. Production Planning & Control:1–11. doi: 10.1080/09537287.2013.782947

Sirkin H, Bradtke T, Lebreton J, Young D (2004) What is globalization doing to your business. https://www.bcgperspectives.com/content/articles/globalization_operations_what_is_globaliza tion_doing_to_your_business/

Spur G (2006) Erscheinungsformen und Modelle technischer Systeme: Ein Beitrag zur theoretischen Begründung der Technikwissenschaften. Parthey, H.; Spur, G.(Hg.): Wissenschaft und Technik in theoretischer Reflexion. Jahrbuch Wissenschaftsforschung:103–130

National Academy of Sciences (1998) Teaching evolution and the nature of science. National Academy Press, Washington, D.C.

Weber J (2008) Überlegungen zu einer theoretischen Fundierung der Logistik in der Betriebswirtschaftslehre. In: Nyhuis P (ed) Beiträge zu einer Theorie der Logistik. Springer-Verlag, Berlin, Heidelberg, pp 43–65

Wiendahl H P (1987) Belastungsorientierte Fertigungssteuerung. Grundlagen, Verfahrensaufbau, Realisierung; 36 Tabellen. Hanser, München

Wiendahl H P, Nyhuis P, Hartmann W (2010) Should CIRP develop a Production Theory? Motivation, Development Path, Framework, 43rd International Conference on Manufacturing Systems, Vienna, 26.-28.05.2010, pp. 3–18

Part II
Individualised Production

Reinhart Poprawe, Frank Piller, Christian Hinke

Individualised production is a concept for designing and aligning all elements of a production system to enable a high level of product variety and dynamics with mass production costs. Product design, production technology and value chains are the key elements within a production system.

Therefore, research in the field of "Individualised Production" within the Cluster of Excellence "Integrative Production Technology for High-wage Countries" is focused on analysing and modelling the optimal combinations and configurations of production system elements and the development of the according processes and technologies.

The emerging Additive Manufacturing (AM) and especially the Selective Laser Melting (SLM) technologies provide great potential for solving the dilemma between scale and scope, i.e. manufacturing products at mass production costs with a maximum fit to customer needs or functional requirements. Due to the complex nature of production systems, the technological potential of AM and especially SLM can only be realised by a holistic comprehension of the complete value creation chain, especially the interdependency between products and production processes.

Therefore, one major objective is the development of a reference architecture for direct, mould-less production systems, applied on SLM production systems, analysing and modelling the interdependencies between the following sub-systems:

- Production technology (structure of SLM machine, manufacturing technology)
- Product design (shape, function, material and mechanical properties)
- Value chain (product programme and business model)

The first chapter of this part discusses the economic effects of AM on the locus of innovation and production. After reviewing some current business models that successfully use AM as a source of value creation, it discusses how AM may enable a more local production by users, supplementing the recent development of an upcoming infrastructure for innovating users and "Makers".

The second chapter of this part gives an overview of production technology, especially on recent results regarding machine concept and process development. Additionally the resulting potential in product design, especially in the field of topology optimization and lattice structures is discussed in this chapter.

Chapter 4
Business Models with Additive Manufacturing—Opportunities and Challenges from the Perspective of Economics and Management

Frank T. Piller, Christian Weller and Robin Kleer

Abstract Technological innovation has frequently been shown to systematically change market structure and value creation. Additive manufacturing (AM), or, colloquially 3D printing, is such a disruptive technology (Berman 2012; Vance 2012). Economic analysis of AM still is scarce and has predominantly focused on production cost or other firm level aspects (e.g., Mellor et al. 2014; Petrovic et al. 2011; Ruffo and Hague 2007), but has neglected the study of AM on value creation and market structure. In this paper, we want to discuss the economic effects of AM on the locus of innovation and production. This is why we first review some current business models that successfully use AM as a source of value creation. Being a potential disruptive influence on market structures, we then discuss how AM may enable a more local production by users, supplementing the recent development of an upcoming infrastructure for innovating users and "Makers".

4.1 Introduction

Recently, it has been highlighted that additive manufacturing (AM) technology has the potential to spark a new industrial revolution by extending the features of conventional production systems (Atzeni and Salmi 2012; Berman 2012; Mellor et al. 2014; The Economist 2011). But AM technology affects market structure beyond direct effects on a single firm's production processes. There is a growing community of "Makers" who develop and share 3D models, sell 3D printed products on marketplaces, and even develop and provide their own 3D printers for home usage (De Jong and de Bruijn 2013; Gershenfeld 2005; Lipson and Kurman 2013). Furthermore, a steadily growing number of 3D printers for home and industrial use extends the scale and scope of manufacturing options. Only two years ago, industry

F.T. Piller (✉) · C. Weller · R. Kleer
TIME Research Area, TIM Group, RWTH Aachen University,
Kackertstraße 7, 52072 Aachen, Germany
e-mail: piller@time.rwth-aachen.de

C. Brecher (ed.), *Advances in Production Technology*,
Lecture Notes in Production Engineering, DOI 10.1007/978-3-319-12304-2_4

analyst Gartner (2012) argued that AM is at its "peak of inflated expectations," noting that the technology is still too immature to satisfy such high expectations. More recently, however, Gartner (2014) predicted that industrial use of AM is likely to reach a level of mainstream adaptation between 2016 and 2020.

AM technology has been in use since the 1980s. In the early phase, the application of AM technology was basically limited to the production of prototypes. The technology's primary goal was to offer an affordable and fast way to receive tangible feedback during the product development process; prototypes were usually not functional (Gibson et al. 2010). Today, prototyping via AM has become a common practice in many firms. The far greater opportunity of AM, however, and the reason behind its current hype, is its promise to replace conventional production technologies for serial manufacturing of components or products ("rapid manufacturing", Gibson et al. 2010). The latter application also bears numerous opportunities for business model innovation.

4.2 Technological Characteristics Driving AM's Economic Impact

Ongoing standardization efforts aim to find a coherent terminology for the various AM technologies in use today. Generally, AM refers to "the process of joining materials to make objects from 3D model data, usually layer upon layer" (ASTM International 2012). There is a variety of different manufacturing processes behind the general term AM. These processes can largely differ in the available choice of materials, build rates, the mechanical properties of the produced parts and other technological constraints. Thus, certain application fields are usually associated with either of these processes. As a result, one cannot refer to *the* AM technology. This is why further economic analysis in this paper aims to generalise some basic principles that characterise AM as a new production technology available for industrial and personal uses.

The main benefit of AM technology is that it enables the flexible production of customized products without cost penalties in manufacturing. It does so by using direct digital manufacturing processes that directly transform 3D data into physical parts, without any need for tools or moulds. Additionally, the layer manufacturing principle can also produce functionally integrated parts in a single production step, hence reducing the need for assembly activities. Thus, AM technology significantly affects the costs of flexibility, individualisation, capital costs, and marginal production costs (Berman 2012; Dolgui and Proth 2010; Koren 2006).

Nonetheless, the opportunities of AM come with a number of limitations: available materials do not always match the characteristics of conventional manufacturing processes, the production throughput speed is rather low, most manufactures still demand an additional surface finish, and common standards for quality control are not established yet (Berman 2012; Gibson et al. 2010). While the former limitations may be of temporary nature, diminishing with technological development, there is a larger inherent threat of AM: In combination with improved 3D-scanning and reverse-engineering capabilities, AM also poses severe risks to the intellectual property rights of product designs (Kurfess and Cass 2014). In the end, AM means digital production, starting with full digital representations of the output. Copying a physical product and converting it into shareable 3D design data might become as easy as copying a printed document or sharing ordinary computer files—similar developments led to disruptive change in the music industry (Wilbanks 2013). The issue of property rights in an age of digital product designs is one of the most severe economic consequences of AM. In the end, we believe it will be the clever design of ecosystems and business models, turning this threat into an opportunity, which will determine the economic potential of AM.

4.3 AM Ecosystem

Economic consequences of AM can hardly be discussed at a single user level. As coined by Jennifer Lawton (president at *MakerBot*), "*3D printing is an ecosystem, not a device*" (Conner 2013). Thus, it is important to develop an understanding of the different elements that constitute this ecosystem which go far beyond sole manufacturing resources and industrial users. Figure 4.1 provides an overview of the components of such an ecosystem.

Though AM is a manufacturing technology, it needs to be considered in the context of digital value chain activities (Brody and Pureswaran 2013; Rayna and Striukova 2014). This is why the ecosystem encompasses activities along a combination of both a conventional manufacturing value chain and a digital value chain of content (product design) creation and distribution. Manufacturing value chains

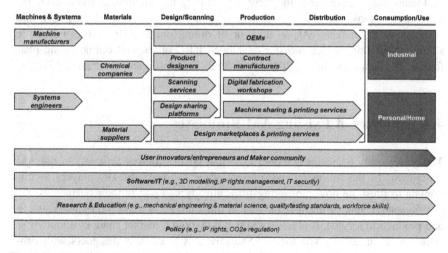

Fig. 4.1 AM ecosystem

frequently include activities related to supply, R&D, production, distribution and the use of a final product (Rayport and Sviokla 1995). Digital value chains differ in regard to its primary object of transactions: it is, by definition, information or digital content (Walters 2012). AM's capability of direct digital manufacturing is frequently highlighted—besides raw materials, it is only the digital product design (CAD) file needed functioning as a universal interface (Berman 2012; Lipson and Kurman 2013; Tuck et al. 2008). Thus, elements of a value chain for digital manufacturing would need to encompass elements such as software, policy (i.e., IP rights), or online services and online 3D design marketplaces.

While most innovation for the manufacturing value chain has been driven by large conventional companies in a BtoB-setting, innovation in the digital value chain has been the result of a growing community of "Makers", i.e. hobbyists, private consumers, and small start-ups interesting in utilizing AM for local manufacture of objects for own use. This community has been very active in developing 3D models, creating an infrastructure for sharing these models digitally in online repositories (like Thingiverse or Google 3D Warehouse), selling 3D printed products on marketplaces, and even developing their own 3D printers for home usage (De Jong and de Bruijn 2013; Gershenfeld 2005; Lipson and Kurman 2013).

We argue that this Maker community has become a kind of "economic lab", experimenting with different designs of value chain and business models, which also provides insight for large scale industrial use of AM. Much of the development of AM innovation in this Maker community has been driven by a mindset of open-source hardware and "Creative Commons" licences. Still, various for-profit businesses emerged successfully from this ecosystem. For example, what started as an open-source project for personal 3D printers (the *RepRap* project), was further developed and commercialized with the '*Makerbot*', a New York based company that became object of a large acquisition by one of the core companies of commercial AM technology (see next section). The upcoming of sustainable business models for AM hence may be similar to the early days of personal computing, where early PC development took place in the "*Homebrew Computer Club*", developing then into commercial PC makers (*Apple* etc.), or of digital music distribution, where file sharing communities like *Napster* developed into commercial online music platforms like *iTunes* (Anderson 2012; Berman 2012; Lipson and Kurman 2013).

4.4 Examples of Existing AM Businesses

Today, a variety of business model exist that cover different activities in the AM ecosystem. In the following, we present some examples of key players in this domain to illustrate how pioneering firms already use AM in their businesses.

Shapeways, one of the first movers in this market, is a 3D model marketplace and production service. It is estimated that they hold a market share of about 70 % (Ponfoort et al. 2014). The idea of *Shapeways* is to connect designers with consumers, thereby collecting a certain service and production fee. *i.materialise* is using a similar business model. *Thingiverse*, on the other hand, is a community-based design sharing platform, operated by *Makerbot* (owned by *Stratasys*). The main idea behind this platform is to promote the use of home 3D printers, in particular Makerbot devices.

FabLabs also aim at promoting the use of 3D printing, however, they are not profit-oriented and work closely together with universities and research centres. These labs provide access to local digital fabrication tools (e.g., 3D printer, laser cutter). *TechShops* commercially provide a similar digital fabrication infrastructure on a pay-by-use basis. *3D Hubs* is a platform to find nearby 3D printers. The idea is to share existing capacity of locally available printers. *3D Hubs* as the match-maker charges a service fee to users. This platform is thus using the advantage of 3D printing as a local production facility.

If files for printing are not downloaded from a design sharing platform like *Thingiverse*, they need to be generated and altered. Already existing objects may be scanned, using a 3D scanner. *NextEngine* or *Makerbot* offer such a product. CAD software, as provided by *Autodesk*, may then be used to edit such files. Alternatively, it can of course be used to generate 3D design files from scratch. Finally, there are, of course, machine manufacturers, such as *3D Systems* or *Stratasys* who offer 3D printers in various price ranges for both, industrial and home use.

Indications of growing market confidence in the sustainability of business models relying on 3D printing offerings include the recently announced acquisition of *Makerbot* by *Stratasys* for US$403 million (Stratasys 2013). Furthermore, dedicated investment funds have been launched that track the performance of the AM sector. Building on AM technology in general and the aforementioned business models in particular, a variety of opportunities for innovation and entrepreneurship arise.

4.5 How AM Facilitates User Innovation and Entrepreneurship

The history of technology taught us that innovation and new business models are frequently developed outside firms' R&D departments (Von Hippel 2005). User-driven innovation appears where problems are directly observed and corresponding solutions are developed. AM facilitates transforming ideas into physical products, and to turn user innovators into manufacturers and entrepreneurs. Thus, user entrepreneurs may become independent of established producers' manufacturing resources to (locally) commercialize their innovations with their own business models.

4.5.1 Local Manufacturing and 3D Printing at Home

A distinctive feature of AM is frequently emphasized in the popular press: its ability to be placed locally next to potential users, up to the point of locating a 3D-printer into a user's home (Berman 2012; De Jong and de Bruijn 2013; The Economist 2011; Vance 2012). Physical products have usually been manufactured at a production site far from the location of end user. For many products fixed costs in conventional production lead to economies of scale. Some products are also simply too difficult to produce or to assemble for a regular user, there is a need for specific knowledge or tools which are costly to get. The downside of this way of producing is typically some kind of missing fit of the final product. Some products are needed "right away", others are produced in a standard setting at the manufacturer while users have a preference for a variety. Moreover, some products require a try-on and rework, again resulting in disutility for the user.

If this disutility overweighs the economies of scale in production, there is scope for local manufacturing at the point of use. This feature is exactly the core of the business model of *3D Hubs*. One of the key characteristics of AM is that it dramatically reduces the benefit of conventional economies of scale. As a result, local manufacturing could become profitable. Anecdotic evidence supports this observation: The price of personal 3D printers has decreased several magnitudes within the last 5 years, leading to a growth in the installed base of this machinery of 50–400 % annually (Wohlers 2013). In addition, an accessible local manufacturing infrastructure based on AM is in the upcoming. Companies like *TechShop* or nonprofit institutions such as *FabLabs* provide local access to AM, comparable to the "copy shop" around the corner. Thus, it is likely that an increasing number of users will direct access to local 3D printing resources in the near future.

4.5.2 User Innovation and AM

Local production may be foremost attractive for innovating users. Past research has shown that users have been the originators of many industrial and consumer products (Von Hippel 2005). Especially when markets are fast-paced or turbulent, these lead users are becoming a major source of innovation. Recent development in IT have lowered the cost for users to innovate: steadily improving design capabilities that advances in computer hardware and software make possible; improved access to easy-to-use development software; and the growth of a steadily richer innovation commons that allows individual users to combine and coordinate their innovation-related efforts via the internet. But there has been a "missing link" (Skinner 1969) in user innovation: manufacturing. Many (lead) users lack the resources and capabilities to turn their inventions into "real" products beyond prototypes, i.e., products with the same properties like industrially manufactured goods. Hence, users often freely revealed their innovations to manufacturers

(Harhoff et al. 2003), benefiting from their capabilities to produce the product in an industrial and stable quality. Manufactures, in turn, benefited from taking up this task by the opportunity to sell these products also to other customers, hence providing a distribution channel for the user invention. For broader development of user innovations, however, this system relied on the availability and willingness of a manufacturer to take up a user innovation.

AM could change this process. Users can turn to advanced AM technologies to produce smaller series of products for themselves and their peers. User innovation then will be supplemented by *user manufacturing*, which we define as the ability of a user to easily turn her design into a physical product. By eliminating the cost for tooling (moulds, cutters) and switching activities, AM allows for an economic manufacturing of low volume, complex designs with little or no cost penalty. AM further enables multiple functionality to be manufactured using a single process, including also secondary materials (like electrical circuits), reducing the need for further assembly for a range of products. In addition, integrated functionality can replace the need for surface coatings and textures (Wohlers 2013). All these characteristics make AM a perfectly suited manufacturing technology for user manufacturers.

4.5.3 User Entrepreneurship and AM

With this production capacity available, user manufacturers may turn into *user entrepreneurs*. Recent research found that innovating (lead) users frequently engage in commercializing their developments (Shah et al. 2012). Accordingly, the term *user entrepreneurship* has been defined as the commercialization of a new product and/or service by an individual or group of individuals who are also innovative users of that product and/or service (Shah and Tripsas 2007). User entrepreneurs experience a need in their life and develop a product or service to address this need, before founding the firm. As a result, user entrepreneurs are distinct from other types of entrepreneurs in that they have personal experience with a product or service that sparked innovative activity and in that they derive benefit through use in addition to financial benefit from commercialization.

The option for local production via AM will also benefit user entrepreneurs. First of all, the sheer opportunity to get access to a flexible manufacturing system without investing in high fixed cost may turn more lead users into user entrepreneurs. In particular, the new product development process can be facilitated when AM is employed. Efforts both in terms of costs and time can be largely reduced with access to local AM resources, while design iterations do not involve cost penalties (no tooling). Once user entrepreneurs started commercializing their products, they may have a competitive advantage against established manufacturers as they obtain better local knowledge on customer demand, allowing them to design products closer to local needs. Especially in a situation where customer demand is heterogeneous and customers place a premium on products fitting exactly to their needs, local producers may outperform established manufacturers of standard goods. The benefits of offering a better product fit may outweigh disadvantages in manufacturing costs due to economies of scales achievable by the established firm with its standard offering. A system of entrepreneurial user manufacturers could have large impact on the market structure in a given industry.

Interestingly, entrepreneurs do not need to acquire their own manufacturing resources. Instead, they might use the existing AM ecosystem and rely on a 3D printing service (like *Shapeways*, as described before) or contract manufacturer to produce their goods—the interface is rather simple: the product's 3D design file. Thus, AM reduces barriers to market entry as fixed costs for production are largely eliminated.

4.6 Conclusions

Concluding, we propose that AM will largely influence the locus of innovation *and* production, enabling the design of new value chains and business models. To achieve economies of scale, many physical products have previously been manufactured far from the site of end use. This can sometimes create high costs for the user due to the lags involved in acquiring something physical that is needed "right away" and "just as I like it". In these cases, AM of physical products at the point of use can make sense even if it comes with high production costs per unit. This market demand, in turn, induces development of on-site manufacturing methods and equipment. Once these are available, they tend to become progressively cheaper and serve larger segments of the market.

However, the future development of AM and its applications are hard to predict, which is mainly caused by the fact that AM is embedded in a large ecosystem with a variety of actors with different capabilities and interests. Users might play a significant role in this ecosystem. They successfully demonstrated their innovating power in the past; now, as they get increasingly more access to local manufacturing resources (formerly, the "missing link"), it is likely that the triad of user innovation, user manufacturing and user entrepreneurship is fuelled. The current rise of a Maker community utilizing, but also developing AM technologies, is a string indicator for this opportunity. In turn, established firms need to rethink their existing business models and adapt a different role in the ecosystem, for example one of a platform operator, marketplace or service provider.

Naturally, there are also opposing drivers, so the question whether production will shift toward a system of local manufacturing is non-trivial: First, under competition, existing manufacturers may react with pricing and/or product enhancements, increasing the appeal of their offerings. Secondly, it has been shown that the strive for economies of scale in a centralized conventional manufacturing system has established a strong and very proven regime that is difficult to break up. Finally, the threshold to engage in own manufacturing may be high for many users. Consider the case of digital photo printing: After a strong rise of home photo printers, the market today has equally divided into decentralized printing kiosks in drugstores and large scale, centralized labs served via the internet. At-home printing of glossy photos however has strongly diminished. Are these transitional adaption effects or structural constraints? Future research has to show.

Acknowledgment The authors would like to thank the German Research Foundation DFG for the kind support within the Cluster of Excellence "Integrative Production Technology for High-Wage Countries.

References

Anderson, C. (2012). Makers: The New Industrial Revolution. Random House.

ASTM International. (2012). ASTM F2792—12a: Standard Terminology for Additive Manufacturing Technologies. ASTM International. Retrieved from www.astm.org/Standards/F2792.htm

Atzeni, E., & Salmi, A. (2012). Economics of Additive Manufacturing for End-usable Metal Parts. The International Journal of Advanced Manufacturing Technology, 62(9–12), 1147–1155.

Berman, B. (2012). 3-D Printing: The New Industrial Revolution. Business Horizons, 55(2), 155–162.

Brody, P., & Pureswaran, V. (2013). The new software-defined supply chain. IBM Institute for Business Value. Retrieved July 12, 2013, from http://public.dhe.ibm.com/common/ssi/ecm/en/gbe03564usen/GBE03564USEN.PDF

Conner, C. (2013). "3D Printing Is An Ecosystem, Not A Device": Jennifer Lawton, MakerBot. Forbes September 13, 2013. Retrieved from http://www.forbes.com/sites/cherylsnappconner/2013/09/13/3d-printing-is-an-ecosystem-not-a-device-jennifer-lawton-makerbot/

De Jong, J. P., & de Bruijn, E. (2013). Innovation lessons from 3-D printing. MIT Sloan Management Review, 54(2), 43–52.

Dolgui, A., & Proth, J.-M. (2010). Supply Chain Engineering: Useful Methods and Techniques. London: Springer.

Gartner. (2012). Hype Cycle for Emerging Technologies August 2012. Retrieved from http://www.gartner.com/newsroom/id/2124315

Gartner. (2014, January 9). Gartner's 2014 Hype Cycle for Emerging Technologies. Retrieved July 17, 2014, from http://www.gartner.com/newsroom/id/2819918

Gershenfeld, N. A. (2005). Fab: The Coming Revolution on Your Desktop. New York, NY: Basic Books.

Gibson, I., Rosen, D. W., & Stucker, B. (2010). Additive Manufacturing Technologies: Rapid Prototyping to Direct Digital Manufacturing. New York, London: Springer.

Harhoff, D., Henkel, J., & von Hippel, E. (2003). Profiting from Voluntary Information Spillovers: How Users Benefit by Freely Revealing their Innovations. Research Policy, 32(10), 1753–1769.

Koren, Y. (2006). General RMS characteristics. Comparison with with dedicated and flexible systems. In A. Dashchenko (Ed.), Reconfigurable Manufacturing Systems: 21st Century Technologies (pp. 27–45). London: Springer.

Kurfess, T., & Cass, W. J. (2014). Rethinking Additive Manufacturing and Intellectual Property Protection. Research-Technology Management, 57(5), 35–42.

Lipson, H., & Kurman, M. (2013). Fabricated: The New World of 3D Printing. Indianapolis, IN: John Wiley & Sons.

Mellor, S., Hao, L., & Zhang, D. (2014). Additive Manufacturing: A Framework for Implementation. International Journal of Production Economics, 149, 194–201.

Petrovic, V., Vicente Haro Gonzalez, J., Jordá Ferrando, O., Delgado Gordillo, J., Ramón Blasco Puchades, J., & Portolés Griñan, L. (2011). Additive layered manufacturing: sectors of industrial application shown through case studies. International Journal of Production Research, 49(4), 1061–1079.

Ponfoort, O., Ambrosius, W., Barten, L., Duivenvoorde, G., van den Hurk, L., Sabirovic, A., & Teunissen, E. (2014). Successfull business models for 3D printing: seizing opportunities with a game changing technology. Utrecht: Berenschot.

Rayna, T., & Striukova, L. (2014). The Impact of 3D Printing Technologies on Business Model Innovation. In P. Benghozi, D. Krob, A. Lonjon, & H. Panetto (Eds.), Digital Enterprise Design & Management (Vol. 261, pp. 119–132). Cham: Springer International Publishing.

Rayport, J. F., & Sviokla, J. J. (1995). Exploiting the virtual value chain. Harvard Business Review, 73(6), 75.

Ruffo, M., & Hague, R. (2007). Cost estimation for rapid manufacturing: simultaneous production of mixed components using laser sintering. Proceedings of the Institution of Mechanical Engineers, Part B: Journal of Engineering Manufacture, 221(11), 1585–1591.

Shah, S. K., & Tripsas, M. (2007). The accidental entrepreneur: The emergent and collective process of user entrepreneurship. Strategic Entrepreneurship Journal, 1(1–2), 123–140.

Shah, S. K., Winston Smith, S., & Reedy, E. J. (2012). Who are user entrepreneurs? Findings on Innovation, Founder Characteristics & Firm Characteristics. Kansas City, MO: Kauffman Foundation.

Skinner, W. (1969). Manufacturing: Missing link in corporate strategy. Harvard Business Review, 47(3), 136–145.

Stratasys. (2013). Press Release: Stratasys to Acquire MakerBot, Merging Two Global 3D Printing Industry Leaders. Retrieved from http://investors.stratasys.com/releasedetail.cfm?ReleaseID= 772534

The Economist. (2011). Print me a Stradivarius. The Economist, February 10, 2011.

Tuck, C. J., Hague, R. J. M., Ruffo, M., Ransley, M., & Adams, P. (2008). Rapid Manufacturing Facilitated Customization. International Journal of Computer Integrated Manufacturing, 21(3), 245–258.

Vance, A. (2012). 3D Printers: Make Whatever You Want. BusinessWeek: Technology April 26, 2012. Retrieved from http://www.businessweek.com/articles/2012-04-26/3d-printers-make-whatever-you-want

Von Hippel, E. (2005). Democratizing Innovation. Cambridge, MA: MIT Press.

Walters, D. (2012). Competition, Collaboration, and Creating Value in the Value Chain. In H. Jodlbauer, J. Olhager, & R. J. Schonberger (Eds.), Modelling Value (pp. 3–36). Heidelberg: Physica-Verlag.

Wilbanks, K. B. (2013). The Challenges of 3D Printing to the Repair-Reconstruction Doctrine in Patent Law. George Mason Law Review, 20(4).

Wohlers, T. T. (2013). Wohlers Report 2013. Fort Collins, CO: Wohlers Associates.

Chapter 5
SLM Production Systems: Recent Developments in Process Development, Machine Concepts and Component Design

Reinhart Poprawe, Christian Hinke, Wilhelm Meiners, Johannes Schrage, Sebastian Bremen and Simon Merkt

Abstract The emerging Additive Manufacturing (AM) and especially the Selective Laser Melting (SLM) technologies provide great potential for solving the dilemma between scale and scope, i.e. manufacturing products at mass production costs with a maximum fit to customer needs or functional requirements. Due to technology intrinsic advantages like one-piece-flow capability and almost infinite freedom of design, Additive Manufacturing was recently even described as "the manufacturing technology that will change the world". Due to the complex nature of production systems, the technological potential of AM and especially SLM can only be realised by a holistic comprehension of the complete value creation chain, especially the interdependency between products and production processes. Therefore this paper aims to give an overview regarding recent research in machine concepts and process development as well as component design which has been carried out within the cluster of excellence "Integrative production technology for high wage countries".

5.1 Introduction

The overall objective of "ICD-A Individualised Production" within the Cluster of Excellence is the resolution of the dichotomy between scale and scope, i.e. manufacturing products at mass production costs with a maximum fit to customer needs or functional requirements (Schleifenbaum 2011).

The emerging Additive Manufacturing (AM) and especially the Selective Laser Melting (SLM) technologies provide great potential for solving this dilemma. With this layer-based technology the most complex products can be manufactured without tools or moulds. The 3D-CAD model gets sliced layer wise for computing the scan tracks of the laser beam. In a first manufacturing step powder material

R. Poprawe · C. Hinke (✉) · W. Meiners · J. Schrage · S. Bremen · S. Merkt
Fraunhofer Institute for Laser Technology ILT, Steinbachstr. 15, 52074 Aachen, Germany
e-mail: christian.hinke@ilt.fraunhofer.de

© The Author(s) 2015

C. Brecher (ed.), *Advances in Production Technology*,
Lecture Notes in Production Engineering, DOI 10.1007/978-3-319-12304-2_5

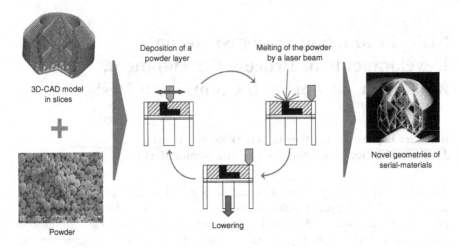

3D-CAD model
in slices

Deposition of a
powder layer

Melting of the powder
by a laser beam

Novel geometries of
serial-materials

Lowering

Powder

Fig. 5.1 Schematic representation of the SLM process

(typically in a range of 25–50 μm) is deposited as a thin layer (typically 30–50 μm) on a substrate plate. According to the computed scan tracks the laser beam melts the powder which is solidified after melting. The substrate plate is lowered and another powder layer is deposited onto the last layer and the powder is melted again to represent the parts geometry (Fig. 5.1). These steps are repeated until almost 100 % dense parts with serial-identical properties are manufactured with the SLM process directly from the 3D-CAD model (Meiners 1999; Over 2003; Schleifenbaum 2008).

Due to technology intrinsic advantages like one-piece-flow capability and almost infinite freedom of design, Additive Manufacturing was recently even described as "the manufacturing technology that will change the world" (Economist 2011) and several international research groups are working on this topic (Gibson 2010; Hopkinson 2005; Lindemann 2006).

AM technology in general and in particular SLM is characterised by a fundamentally different relation of cost, lot size and product complexity compared to conventional manufacturing processes (Fig. 5.2). There is no increase of costs for small lot sizes (in contrast to mould-based technologies) and no increase of costs for shape complexity (in contrast to subtractive technologies).

For conventional manufacturing technologies such as die casting the piece cost depends on the lot size. For increasing lot sizes the piece costs are decreasing due to economies-of-scale. Because lot depended fixed costs (e.g. tooling costs) are very low for AM, AM enables the economic production of parts in small lot sizes (Individualisation for free). Innovative business models such as customer co-creation can be implemented using this advantage of AM technologies (Fig. 5.2 left).

The more complex a product is, the piece cost for manufacturing increase. For AM this relation is not applicable. The nearly unlimited geometric freedom that is offered through AM makes the piece cost almost independent from product complexity. In some cases manufacturing costs can even decrease due to lower build-up

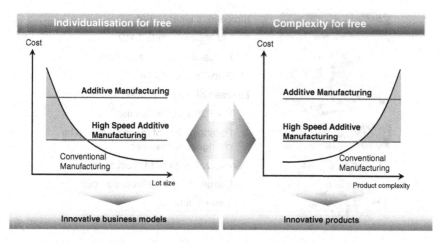

Fig. 5.2 Innovative business models and innovative products enabled by Additive Manufacturing

volumes of optimized products with high geometric complexity. Topology optimisation is one design approach to save weight while functionally adapting the product design to predefined load cases (Huang 2007). These different relations between piece cost and product complexity offer a unique capability for AM to manufacture innovative products perfectly adapted to the specific technological requirements through the integration of lattice structures (Fig. 5.2 right).

Due to the complex nature of production systems, the technological potential of AM and especially SLM can only be realised by a holistic comprehension of the complete value creation chain, especially the interdependency between products and production processes.

Therefore this paper aims to give an overview regarding recent research in machine concepts and process development as well as component design which has been carried out within the cluster of excellence "Integrative production technology for high wage countries".

5.2 SLM Machine Concepts

State-of-the-art SLM machines are typically equipped with a 400 W laser beam source and a build space of $250 \times 250 \times 300$ mm^3. There are different ways to increase the productivity of SLM machines in terms of process build rate. In general the productivity can be increase by the following measures:

Increase of laser power to increase the scanning speed, use bigger layer thickness and bigger beam diameter to increase the build-up speed.

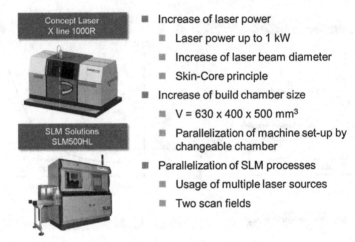

Fig. 5.3 SLM machine concepts for increasing the productivity

By increasing the build volume more parts can be manufactured in one build job. Hence, the number of necessary machine set-up and part removal procedures can be reduced.

Another method to increase the productivity is the parallelisation of the SLM process by using multiple laser beam sources and multi laser-scanning-systems in one machine. Either the build area can be multiplied or one build space can be processed by multi lasers and scanning-systems at the time.

The SLM machine X-line 1000R by Concept Lasers that was presented at the Euromold (2012) comprises a big build space of $630 \times 400 \times 500$ mm^3 that can be processed with one laser-scanning-system that is moved above the build platform (Fig. 5.3). This machine is equipped with two process chambers that are rotatable. One process chamber can be prepared for the next build job while in the second one the SLM process can take place (Concept Laser 2014).

Another example of a SLM machine with an increased build volume is the SLM machine SLM500HL by SLM Solutions. The build volume is $500 \times 280 \times 335$ mm^3. The process chamber can be moved inside and out of the machine. The preparation of the build space can be done outside of the process chamber. The SLM machine SLM500HL can be equipped with up to 4 laser beam sources with that the build are can be processed at the same time. Furthermore the so called skin-core build strategy can be performed. The beam with a small diameter ($d_S = 80$ μm) and up to a laser power of 400 W is used to build up the outer shape of the parts (skin). A laser beam with big beam diameter ($d_S = 700$ μm) and laser power up to 1 kW is used to build the inner core of the part (core). So the build up speed can be elevated while maintaining the part accuracy (SLM Solutions 2014).

5.2.1 Valuation Method for SLM Machine Concepts

For the comparison and the evaluation of different SLM machine concepts not only the investment costs of the machine have to be taken into account but also the costs that are caused during the whole utilization time. Therefore a suitable valuation method has to be found.

One approach is to analyse the life cycle costs (such as the machine price, maintenance costs and energy costs) and the cost of the life cycle performance (such as "laser-on-time" to "laser-off time").

In typical cost models for SLM investment evaluation the machine hourly rate is used to calculate the price for the SLM manufactured part. Furthermore, cost and time factors for preparation and follow steps may be included in the calculation (Rickenbacher 2013).

In a life cycle cost analysis, including the identification of cost drivers is carried out for additively manufactured components (Lindemann 2012). As the main cost driver, the machine costs (73 %) and material costs (11 %) were followed by the cost of any demand (7 %) and preparatory work (4 %) are identified.

In none of those cost models the influence of the SLM machine concept (build space, laser power, etc.) is analysed in detail. The cost drivers of those SLM machines are undiscovered.

A life cycle cost based model is developed to create a possibility to compare different SLM machine concepts.

In Fig. 5.4 the costs for molten powder material volume (€/cm^3) are shown as a function of the total laser power used in the SLM process. The allocation of the total costs can be divided into the investment costs (machine costs, costs for laser) and variable costs (powder material, energy costs, gas costs etc.). In this case the operating time of the machine is seven years at a capacity utilization of 70 %. The

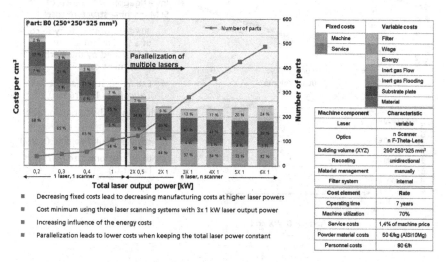

Fig. 5.4 Influence of total laser power on SLM manufactured part costs

build space is considered with $250 \times 250 \times 325$ mm^3 and the part to be manu-factured is a space-filling solid cuboid of the dimension of $250 \times 250 \times 325$ mm^3. Here only one laser-scanning system is used up to 1 kW total laser power. At total laser power >1 kW the SLM process is parallelised by integrating multiple laser scanner systems in one SLM machine.

The findings are: The costs of molten material are reduced when using higher total laser power due to increasing productivity. The minimum costs are found at a total laser power of 3 kW, which is provided by 3×1 kW laser and scanner systems. With further increase of the total laser power (>3 kW) and hence inte-grating more laser beam sources and more laser-scanning systems the increasing energy costs lead to an increase in total manufacturing costs.

The cost of the SLM machine itself represents the largest share of the total part manufacturing costs. In Fig. 5.5 the effects of the machine build volume in relation to the machine costs are illustrated. At a same physical height the enlargement of the space in the lateral plane (X and Y direction) has a stronger influence on the machine costs, as a space, which is increased by increasing the height (Z direction). Up to a ratio of three for the Z-height to its X–Y-area cost savings can be identified. At ratios greater than 3, only minor changes in the costs are observed.

5.2.2 SLM Machine Concept Parallelization

Within the framework of the Cluster of Excellence a multi-scanner-SLM machine is designed and implemented. In this system two lasers and two scanners are

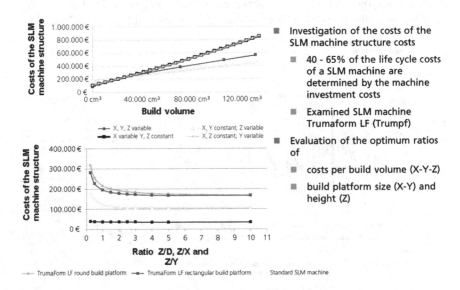

Fig. 5.5 Impact of build volume on the SLM manufactured part costs

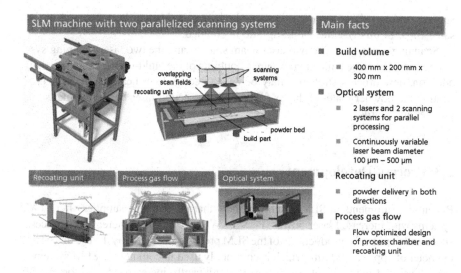

Fig. 5.6 SLM machine concept of parallelization

integrated. These two scanners can be positioned to each other that both either scan two fields on its own (double-sized build space) or one scan field is processed with two scanners simultaneously (Fig. 5.6).

With these multi-scanner systems new scanning strategies can be developed and implemented (Fig. 5.7).

Scanning strategy 1: Both scan fields are positioned next to each other with a slight overlap. This results in a doubling of the build area. By using two laser beam

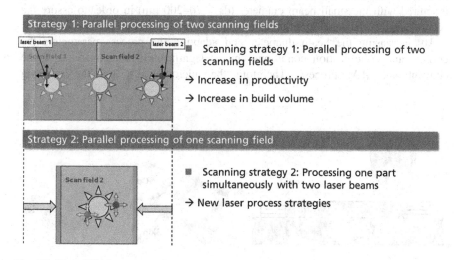

Fig. 5.7 New SLM laser processing strategies with two lasers and two scanning systems

sources and two laser-scanning systems both scan fields can be processed at the same time. In this case the build-up rate is doubled.

Scanning strategy 2: The two laser beam sources and the two laser-scanning systems expose the same build area. Again, a doubling of the build-up rate is achieved. In addition, new process strategies may be developed: A laser beam is used for preheating the powder material that is followed by a second laser beam that melts the powder afterwards.

5.3 Process Development

Recent developments in SLM machines show that the machine supplier offer SLM systems with increased laser power ($P_L \leq 1$ kW). The aim is to increase the process speed and thereby the productivity of the SLM process. However, by the use of a beam diameter of approx. 100 μm, which is commonly used in commercial SLM systems, the intensity at the point of processing is significantly increased due to the use of increased laser power. This effect results in spattering and evaporation of material and therefore in an unstable and not reproducible SLM process. For this reason the beam diameter has to be increased in order to lower the intensity in the processing zone. In this case the melt pool is increased and the surface roughness of the manufactured part is negatively influenced. To avoid these problems the so-called skin-core strategy (Fig. 5.8) is used, whereby the part is divided into an inner core and an outer skin (Schleifenbaum 2010). Different process parameters and focus diameters can be designated to each area. The core does not have strict limitations or requirements concerning the accuracy and detail resolution. Therefore, the core area can be processed with an increased beam diameter (ds = 400–1000 μm) and an increased laser power; thus resulting in an increased productivity. In contrast, the skin area is manufactured with the small beam diameter (ds = 70–200 μm) in order to assure the accuracy and surface quality of the part.

By increasing the beam diameter and adapting the process parameters the cooling and solidification conditions change significantly in comparison to the conventional SLM process. Therefore the microstructure and the resulting

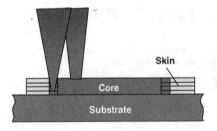
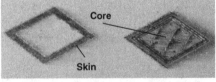

Fig. 5.8 Principle skin-core strategy

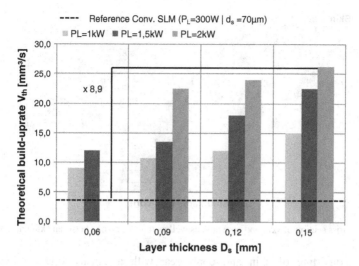

Fig. 5.9 Theoretical build up rate according to layer thickness and laser power processing the tool steel 1.2709

mechanical properties have to be investigated in detail. These investigations are done for the maraging tool steel 1.2709 within the Cluster of excellence.

The first step is to investigate process parameters on cubic test samples which have a averaged density of ≥99.5 %. Therefore a SLM machine setup with a laser beam diameter of d_s = 80 μm (Gaussian beam profile) and d_s = 728 μm (Top-hat beam profile) is used. The results for the achieved theoretical build-up rate which is calculated by the product of hatch distance, layer thickness and scanning velocity is illustrated in Fig. 5.9. It can be observed that by the increase of the laser power from 300 W up to P_L = 1 kW and an adaption of the process parameters layer thickness and scanning velocity the theoretical build up rate can be increased from 3 mm^3/s to 15 mm^3/s. A further increase of the laser power up to P_L = 2 kW results in an increase of the theoretical build-up rate to 26 mm^3/s (factor 8,9). These investigations show that it is possible with the use of increased laser power up to P_L = 2 KW to heighten the theoretical build-up rate and thereby the productivity significantly.

After the investigation of process parameters the microstructure for the manufactured test samples is investigated. Especially the transition zone between skin and core area has to be investigated in order to assure a metallurgical bonding between skin and core area. Due to the different layer thicknesses ($D_{s,skin}$ = 30 μm, $D_{s,core}$ = 60–150 μm) the scanning strategy has to be adapted.

Figure 5.10 shows etched cross section of the skin and core area as well as the transition zone between skin and core area. It can be observed that due to the use of the increased beam diameter the melt pool size is significantly increased. In addition the transition zone between skin and core area shows no defects and metallurgical bonding. As a result it can be noted that by the use of the skin-core strategy, test

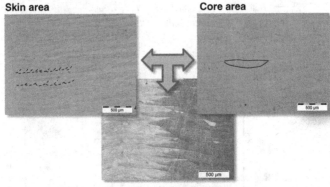

Fig. 5.10 Cross section of skin and core area as well as transition zone between skin and core area

samples, consisting of skin and core area, with a density ≥99.5 % can be manufactured.

In order to investigate the properties of the additive manufactured parts mechanical tests are carried out. Therefore tensile test specimens with the dimension B5x25 according to DIN50125 are manufactured and afterwards heat treated. For each parameter five tensile test specimens are manufactured and the averaged values for tensile strength Rm, yield strength Rp0,2 and the breaking elongation are measured.

Figure 5.11 shows the results of the tensile tests according to the employed process parameters. It can be observed that for conventional SLM using 300 W laser power at 30 µm layer thickness a tensile strength Rm = 1060 N/mm^2 is achieved. By the use of the heat treatment the tensile strength is increased up to Rm = 1890 N/mm^2. In contrast the heat treatment leads to an reduced breaking

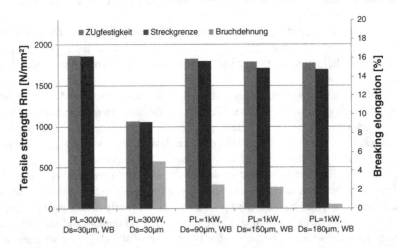

Fig. 5.11 Results tensile test according to process parameters

elongation (5 % → 1.32 %). These results are used as reference for the tensile test specimen manufactured by the use of increased laser power. The results in Fig. 5.4 show that by an increase of the laser power up to PL = 1 kW and an adaption of process parameters as well as the use of a heat treatment a tensile strength of Rm = 1824 N/mm^2 is achieved. In addition the breaking elongation using PL = 1 kW shows a value of 2.53 %. A further increase of the laser power up to PL = 1.5 kW leads to tensile strength of 1790 N/mm^2 with a breaking elongation of 2.24 %. These results show that the mechanical properties for the maraging steel 1.2709, processed with a laser power up to 1.5 kW, lead to mechanical properties which are in the same range as the properties for conventional SLM (PL = 300 W).

5.4 Functional Adapted Component Design

As explained in Fig. 5.2 Additive Manufacturing in general and especially SLM provides a great potential for innovative business models and innovative products or components. Due to technology intrinsic advantages like "Individualisation for free" and "Complexity for free", SLM is the technology of choice for the production of functional adapted products or components in small or medium lot sizes.

In order to raise this potential, design methods has to be adapted to the potential of Additive Manufacturing (e.g. topology optimisation) or even new design methods has to be developed (e.g. lattice structures).

The following chapter shows recent results in the field of topology optimisation and lattice structure design.

5.4.1 Topology Optimisation and SLM

In contrast to subtractive manufacturing methods like machining, the main cost driver of the SLM process is the process time needed to generate a certain amount of part volume. As a consequence, reducing the part volume to the lowest amount needed to absorb the forces of the use case, is an important factor to increase the productivity of the SLM process. Topology optimisation is an instrument to design load adapted parts based on a FEM-analysis. The load cases including forces and clamping need to be very clear to get the best results possible. In an iterative process the topology optimisation algorithm calculates the stress level of each FEM element. Elements with low stresses are deleted until the optimisation objective/criterion (e.g. weight fraction) is reached. The topology optimisation usually results in very complex parts with 3D freeform surfaces and hollow and filigree structures. The commonly used way to fabricate these part designs is a reconstruction considering process restrictions of conventional manufacturing methods like casting or machining, resulting in a lower weight reduction. SLM opens opportunities to fabricate complex optimisation designs without any adjustments after optimisation.

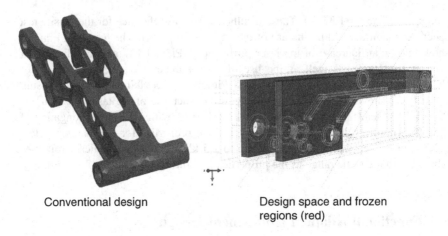

Conventional design   Design space and frozen
                                       regions (red)

Fig. 5.12 Kinematics lever of a business class seat

The reduction of weight is an important factor in aerospace industry (Rehme 2009). Fuel consumption is mainly determined by the weight of the aircraft. An aircraft seat manufacturer is investigating the opportunities to save weight in their business class seats through SLM. One part of the seat assembly, a kinematics lever, was selected to investigate the potential of the direct fabrication of topology optimisation results via SLM (Fig. 5.12).

In a first step the maximum design space and connecting interfaces to other parts in the assembly were defined (Fig. 5.12) to guarantee the fit of the optimisation result to the seat assembly. Interfacing regions are determined as frozen regions, which are not part of the design space for optimisation. The kinematics lever is dynamically loaded if the passenger takes the sleeping position. Current topology optimisation software is limited to static load cases. Therefore the dynamic load case is simplified to five static load cases, which consider the maximum forces at different times of the dynamic seat movement. Material input for the optimisation is based on an aluminium alloy (7075) which is commonly used in aerospace industry: material density: 2810 kg/m^3, E Modulus: 70.000 MPa, Yield Strength: 410 MPa, Ultimate Tensile Strength: 583 MPa and Poisson's Ratio: 0.33. The objective criterion of the optimisation is a volume fraction of 15 % of the design space. The part is optimized regarding stiffness. In Fig. 5.13 the optimisation result as a mesh structure and a FEM analysis for verification of the structure are shown.

The maximum stress is approx. 300 MPa, which is below the limit of Yield Strength of 410 MPa. Before the manufacturing of the optimisation result via SLM, the surfaces get smoothened to improve the optical appearance of the part. Compared to the conventional part (90 g) a weight reduction of approx. 15 % (final weight 77 g, Fig. 5.14) was achieved. For a series production of this part further improvements to increase the productivity of the process are needed.

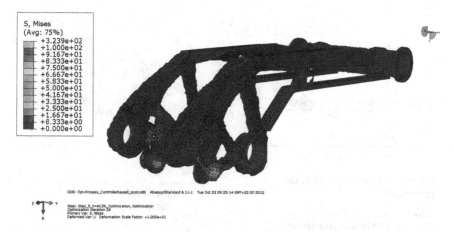

Fig. 5.13 Mesh structure of optimisation result including stress distribution

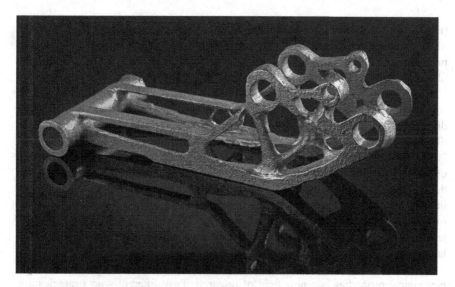

Fig. 5.14 Final light-weight part manufactured by SLM

5.4.2 Functional Adapted Lattice Structures and SLM

The almost unlimited freedom of design offered by SLM provides new opportunities in light-weight design through lattice structures. Due to unique properties of lattice structures (good stiffness to weight ratio, great energy absorption, etc.) and their low volume, the integration of functional adapted lattice structures in functional parts is a promising approach for using the full technology potential of SLM

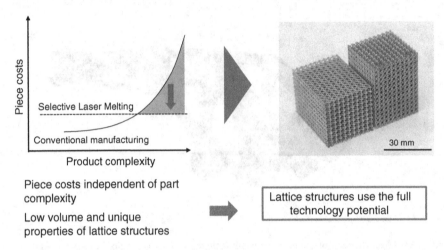

Fig. 5.15 Complexity-for-free offers great opportunities through lattice structures

(Fig. 5.15). Compared to conventional manufacturing technologies, piece costs of SLM parts are independent of part complexity and the main cost driver is the process time (correlates with part volume). Lattice structures can reduce the amount of part volume and host unique properties.

Three main challenges need to be solved to make lattice structures a real option for the use in functional parts in different industries. The mechanical properties of different lattice structure types were studied by several researchers (Löber 2011; Shen 2010, 2012; Yan 2012; Rehme 2009; Gümrück 2013; Smith 2013; Ushijima 2011). Nevertheless, there is no comprehensive collection of mechanical properties of lattice structures under compressive, tensile, shear and dynamic load. Also the deformation and failure mechanisms are not studied sufficiently. A relatively new field of research is the influence of different scan parameters/strategies on the mechanical properties. To reach the overall objective of our research these challenges need to be overcome to design functional adapted parts with integrated lattice structures (Fig. 5.16).

As said before, the correlation between process parameters/scan strategy and mechanical properties is a new field of research. Two different scan strategies are commonly used for the fabrication of lattice structures by SLM: Contour-Hatch scan strategy and Pointlike exposure (Fig. 5.17).

Contour-Hatch scan strategy is the most commonly used scan strategy, which causes many scan vectors and jumps between scan vectors, resulting in a high amount of scanner delays. Pointlike exposure strategy reduces the complex geometry to a set of points of exposure and less jumps and scanner delays are caused. To investigate the influence of the two scan strategies on the geometry of the lattice structures, different types of f2ccz structures were manufactured. The material used in this study was stainless steel 316L (1.4404) from TLS. The parameters were iteratively optimized regarding a low geometric deviation from the

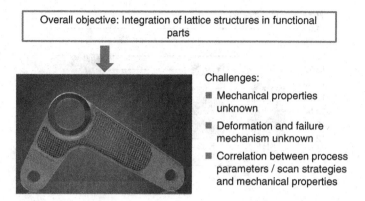

Fig. 5.16 A new way of designing functional parts by the integration of lattice structures

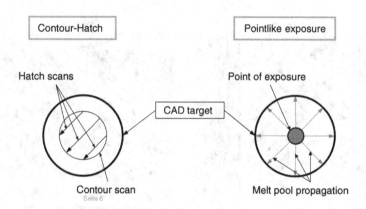

Fig. 5.17 Commonly used scan strategies for the fabrication of lattice structures

CAD model. A measurement of the relative density of the lattice structures by archimedean density measurement was performed. The relative density is the filling degree of the structure and can be used to determine geometric deviations of the structure. Three different kinds of Contour-Hatch parameters (Laser power: 100–130 W, scan speed: 700–900 mm/s) and one parameter set for Pointlike exposure (Laser power: 182 W) were investigated. Figure 5.18 shows the deviations of the relative density to the CAD model target for the investigated parameters.

For Pointlike exposure strategy the relative density is 4 % higher than the CAD model target. All in all, the CAD model target can be reached with low deviations. To further investigate the geometry lattice structures were investigated by micro CT measurement. Figure 5.19 shows a reconstruction based on these CT images.

Lattice structures manufactured by Contour-Hatch scan strategy show no visible build-up errors and vertical and diagonal struts have the same diameter. In contrast pointlike exposure strategy show light contractions at knots and deviations between vertical and diagonal strut diameter.

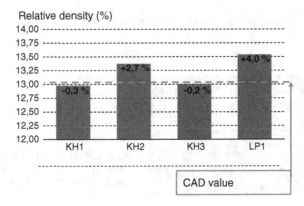

Fig. 5.18 Deviations of relative density to the CAD model target

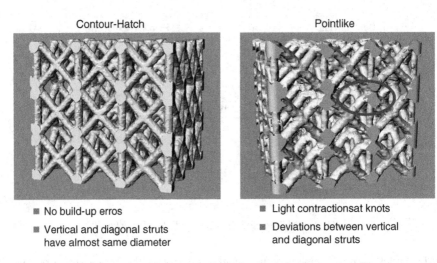

Contour-Hatch   Pointlike

- No build-up erros
- Vertical and diagonal struts have almost same diameter

- Light contractionsat knots
- Deviations between vertical and diagonal struts

Fig. 5.19 Mirco CT reconstructions to investigate the dimensional accuracy of lattice structures

Acknowledgment The authors would like to thank the German Research Foundation DFG for the kind support within the Cluster of Excellence "Integrative Production Technology for High-Wage Countries.

References

Concept Laser (2014) Data sheet X line 1000R. http://platforms.monash.edu/mcam/images/stories/Concept/xline_1000.pdf Accessed 02.10.2014

Economist (2011) Cover Story, 12.2.2011

Gibson I et al (2010) Additive Manufacturing Technologies. Rapid Prototyping to Direct Digital Manufacturing, Springer, Heidelberg

Gümrük R, Mines R A W (2013) Compressive behaviour of stainless steel micro-lattice structures, International Journal of Mechanical Sciences, 2013

Hopkinson N et al (eds) (2005) Rapid Manufacturing: An Industrial Revolution for the Digital Age. Wiley

Huang X, Xie M (2010) Evolutionary Topology Optimisation of Continuum Structures: Methods and Applications. Wiley

Lindemann U (eds) (2006) Individualisierte Produkte - Komplexität beherrschen in Entwicklung und Produktion, Springer, Heidelberg

Lindemann C et al (2012) Analyzing Product Lifecycle Costs for a Better Understanding of Cost Drivers in Additive Manufacturing. In: Proceedings of "International Solid Freeform Fabrication Symposium 2012", Austin Texas

Löber L, Klemm D, Kühn U, Eckert, J (2011) Rapid Manufacturing of Cellular Structures of Steel or Titaniumalumide, MSF (Materials Science Forum), Vol. 690, 2011

Meiners W (1999) Direktes Selektives Lasersintern einkomponentiger metallischer Werkstoffe. Dissertation, RWTH Aachen

Over C (2003) Generative Fertigung von Bauteilen aus Werkzeugstahl X38CrMoV5-1 und Titan TiAL6V4 mit „Selective Laser Melting". Dissertation, RWTH Aachen

Rehme O (2009) Cellular Design for Laser Freeform Fabrication, PhD Thesis, Laser Zentrum Nord, Hamburg

Rickenbacher L (2013) An integrated cost-model for selective laser melting. Rapid Prototyping Journal, Vol. 19 Number 3 2013, pp. 208–214

Schleifenbaum H, Meiners W, Wissenbach K (2008) Towards Rapid Manufacturing for series production: an ongoing process report on increasing the build rate of Selective Laser Melting (SLM). International Conference on Rapid Prototyping & Rapid Tooling & Rapid Manufacturing, Berlin, Germany

Schleifenbaum H et al (2010) Individualized production by means of high power Selective Laser Melting. CIRP Journal of Manufacturing Science and Technology, vol 2 (3), pp. 161–169

Schleifenbaum H et al (2011) Werkzeuglose Produktionstechnologien für individualisierte Produkte. In: Brecher C (eds) Integrative Produktionstechnik für Hochlohnländer, Springer

Shen Y, McKown S, Tsopanos S, Sutcliffe C J, Mines R A W, Cantwell W J (2010) The Mechanical Properties of Sandwich Structures Based on Metal Lattice Architectures, Journal of Sandwich Structures and Materials, Vol. 12, 2010

Shen Y, Cantwell W J, Mines R A W, Ushijima K (2012) The Properties of Lattice Structures Manufactured Using Selective Laser Melting, AMR (Advanced Materials Research), Vol. 445, pp. 386–391, 2012

SLM Solutions (2014) Data sheet SLM500HL. http://www.stage.slm-solutions.com/download.php?f=8277396e9b97045edbb7ef680e3ada56 Accessed 02.10.2014

Smith M, Guan Z, Cantwell W J (2013) Finite element modelling of the compressive response of lattice structures manufactured using the selective laser melting technique, International Journal of Mechanical Sciences, 2013

Ushijima K, Cantwell W J, Mines R A W, Tsopanos S, Smith M (2011) An investigation into the compressive properties of stainless steel micro-lattice structures, Journal of Sandwich Structures and Materials, 2011

Yan C, Hao L, Hussein A, Raymont D (2012) Evaluations of cellular lattice structures manufactured using selective laser melting, International Journal of Machine Tools and Manufacture, Vol 62, pp. 32–38, 2012

Part III
Virtual Production Systems

Gerhard Hirt, Markus Bambach, Wolfgang Bleck,
Ulrich Prahl and Wolfgang Schulz

Computational methods have radically changed the way engineers design materials, products and manufacturing processes. Numerical simulations are used to save resources, e.g. by reducing the need for expensive experiments, to predict and optimize properties that cannot be measured directly, such as the microstructure of a material in a production process, and to explore new processes and parameter ranges in known processes that are not easily accessible experimentally, e.g. if this would require expensive new equipment. The industrial needs have been a steady driver for innovation in the numerical simulation of manufacturing engineering processes. In the automotive industry, for instance, it has become common practice to design metal parts and the corresponding manufacturing processes 'virtually' before building expensive tool sets. In materials science and engineering, the computer-aided development of new materials has started to replace the 'alchemistic' way of materials design.

With the availability of vast computing power, the development of parallel processing and robust numerical methods, it seems that not only individual manufacturing processes could be simulated but that the entire processing chain of a product 'from the cradle to the grave' could be designed virtually. This scenario is currently being pursued in the emerging field of 'integrated computational materials engineering' (ICME), which is an integrative approach for developing products, materials and the corresponding manufacturing processes by coupling of simulations across physical length scales and along the manufacturing process chain.

Matured numerical simulation, as well as experimental diagnosis in manufacturing and materials engineering create data sets that are difficult to interpret. The data sets are sparse in multi-dimensional parameter space since their generation is expensive. Using standard methods for data manipulation, like optimization criteria, supporting decision-making is difficult since the data are often discrete. Also, immense data streams created in the shop floor by sensors and computerized quality management are not well suited since they tend to be unnecessarily dense. Model reduction, meta-modelling and visualization approaches are hence needed to prepare, explore and manipulate the raw data sets emerging from manufacturing metrology and virtual production.

This session deals with state-of-the-art methods of virtual production systems, which enable the planning of manufacturing and production processes, the handling of raw data sets and the development of new materials. Two key issues are addressed:

The paper "Meta-modelling techniques towards virtual production intelligence" addresses the problem of handling data sets and generation of information by means of meta-modelling techniques. In the example of laser sheet metal cutting it is shown how meta-models can be used to reduce complexity and allow decision-making.

The contribution "Designing new forging steels by ICMPE" envisions the next development step of ICME by achieving coupling to production engineering. The benefit of the resulting field of Integrated Computational Materials and Production Engineering (ICMPE) is shown with the aid of newly developed forging steels whose microstructure is designed by controlling precipitation kinetics and structural size using closely interacting alloying and processing concepts.

Chapter 6
Meta-Modelling Techniques Towards Virtual Production Intelligence

Wolfgang Schulz and Toufik Al Khawli

Abstract Decision making for competitive production in high-wage countries is a daily challenge where rational and irrational methods are used. The design of decision making processes is an intriguing, discipline spanning science. However, there are gaps in understanding the impact of the known mathematical and procedural methods on the usage of rational choice theory. Following Benjamin Franklin's rule for decision making formulated in London 1772, he called "Prudential Algebra" with the meaning of prudential reasons, one of the major ingredients of Meta-Modelling can be identified finally leading to one algebraic value labelling the results (criteria settings) of alternative decisions (parameter settings). This work describes the advances in Meta-Modelling techniques applied to multidimensional and multi-criterial optimization in laser processing, e.g. sheet metal cutting, including the generation of fast and frugal Meta-Models with controlled error based on model reduction in mathematical physical or numerical model reduction. Reduced Models are derived to avoid any unnecessary complexity. The advances of the Meta-Modelling technique are based on three main concepts: (i) classification methods that decomposes the space of process parameters into feasible and non-feasible regions facilitating optimization, or monotone regions (ii) smart sampling methods for faster generation of a Meta-Model, and (iii) a method for multi-dimensional interpolation using a radial basis function network continuously mapping the discrete, multi-dimensional sampling set that contains the process parameters as well as the quality criteria. Both, model reduction and optimization on a multi-dimensional parameter space are improved by exploring the data mapping within an advancing "Cockpit" for Virtual Production Intelligence.

W. Schulz (✉)
Fraunhofer Institute for Laser Technology ILT,
Steinbachstr. 15, 52074 Aachen, Germany
e-mail: wolfgang.schulz@ilt.fraunhofer.de

T.A. Khawli
Nonlinear Dynamics of Laser Processing of RWTH Aachen,
Steinbachstr. 15, 52074 Aachen, Germany
e-mail: toufik.al.khawli@ilt.fraunhofer.de

C. Brecher (ed.), *Advances in Production Technology*,
Lecture Notes in Production Engineering, DOI 10.1007/978-3-319-12304-2_6

6.1 Introduction

Routes of Application

At least two routes of direct application are enabled actually by Meta-Modelling, namely, decision making and evaluation of description models. While calculating multi-objective weighted criteria resulting in one algebraic value applies for decision making, multi-parameter exploration for the values of one selected criterion is used for evaluation of the mathematical model which was used to generate the Meta-Model.

Visual exploration and Dimensionality Reduction

More sophisticated usage of Meta-Modelling deals with visual exploration and data manipulation like dimensionality reduction. Tools for viewing multidimensional data (Asimov 2011). are well known from iterature. Visual Exploration of High Dimensional Scalar Functions (Gerber 2010) today focusses on steepest-gradient representation on a global support, also called Morse-Smale Complex. The scalar function represents the value of the criterion as function of the different parameters. As result, at least one trace of steepest gradient is visualized connecting an optimum with a minimum of the scalar function. Typically, the global optimum performance of the system, which is represented by a specific point in the parameter space, can be traced back on different traces corresponding to the different minima. These trace can be followed visually through the high-dimensional parameter space revealing the technical parameters or physical reasons for any deviation from the optimum performance.

Analytical methods for dimensionality reduction, e.g. the well-known Buckingham Π-Theorem (Buckingham 1914), are applied since 100 years for determination of the dimensionality as well as the possible dimensionless groups of parameters. Buckingham's ideas can be transferred to data models. As result, methods for estimating the dimension of data models (Schulz 1978), dimensionality reduction of data models as well as identification of suitable data representations (Belkin 2003) are developed.

Value chain and discrete to continuous support

The value chain of Meta-Modelling related to decision making enables the benefit rating of alternative decisions based on improvements as result of iterative design optimization including model prediction and experimental trial. One essential contribution of Meta-Modelling is to overcome the drawback of experimental trials generating sparse data in high-dimensional parameter space. Models from mathematical physics are intended to give criterion data for parameter values dense enough for successful Meta-Modeling. Interpolation in Meta-Modelling changes the discrete support of parameters (sampling data) to a continuous support. As result of the continuous support, rigorous mathematical methods for data manipulation are applicable generating virtual propositions

Resolve the dichotomy of cybernetic and deterministic approaches

Meta-Modelling can be seen as a route to resolve the separation between cybernetic/empirical and deterministic/rigorous approaches and to bring them together as well as making use of the advantages of both. The rigorous methods involved in Meta-Modelling may introduce heuristic elements into empirical approaches and the analysis of data, e.g. sensitivity measures, may reveal the sound basis of empirical findings and give hints to reduce the dimension of Meta-Models or at least to partially estimate the structure of the solution not obvious from the underlying experimental/numerical data or mathematical equations.

6.2 Meta-Modelling Methods

In order to gain a better insight and improve the quality of the process, the procedure of conceptual design is applied. The conceptual design is defined as creating new innovative concept from simulation data (Currie 2005). It allows creating and extracting specific rules that potentially explain complex processes depending on industrial needs.

Before applying this concept, the developers are validating their model by performing one single simulation run (s. Application: sheet metal drilling) fitting one model parameter to experimental evidence. This approach requires sound phenomenological insight. Instead of fitting the model parameter to experimental evidence multi-physics, complex numerical calculations can be used to fit the empirical parameters of the reduced model. This requires a lot of scientific modelling effort in order to achieve good results that could be comparable to real life experimental investigation.

Once the model is validated and good results are achieved, the conceptual design analysis is then possible either to understand the complexity of the process, optimize it, or detect dependencies. The conceptual design analysis is based on simulations that are performed on different parameter settings within the full design space. This allows for a complete overview of the solution properties that contribute well to the design optimization processes (Auerbach et al. 2011). However the challenge rises when either the number of parameters increases or the time required for each single simulation grows.

Theses drawbacks can be overcome by the development of fast approximation models which are called metamodels. These metamodels mimic the real behavior of the simulation model by considering only the input output relationship in a simpler manner than the full simulation (Reinhard 2014). Although the metamodel is not perfectly accurate like the simulation model, yet it is still possible to analyze the process with decreased time constraints since the developer is looking for tendencies or patterns rather than values. This allows analyzing the simulation model much faster with controlled accuracy.

Meta-modelling techniques rely on generating and selecting the appropriate model for different processes. They basically consist of three fundamental steps: (1)

the creation and extraction of simulation data (sampling), (2) the mapping of the discrete sampling points in a continuous relationship (interpolation), and (3) visualization and user interaction of this continuous mapping (exploration)

1. Sampling
2. Interpolation
3. Exploration.

6.2.1 Sampling

Sampling is concerned with the selection of discrete data sets that contain both input and output of a process in order to estimate or extract characteristics or dependencies. The procedure to efficiently sampling the parameter space is addressed by many Design of Experiments (DOE) techniques. A survey on DOE methods focusing on likelihood methods can be found in the contribution of Ferrari et al. (Ferrari 2013). The basic form is the Factorial Designs (FD) where data is collected for all possible combinations of different predefined sampling levels of the full the parameter space (Box and Hunter 1978).

However for high dimensional parameter space, the size of FD data set increases exponentially with the number of parameters considered. This leads to the well-known term "Curse of dimensionality" that was defined by Bellman (Bellman 1957) and unmanageable number of runs should be conducted to sample the parameter space adequately. When the simulation runs are time consuming, these FD design could be inefficient or even inappropriate to be applied on simulation models (Kleijnen 1957).

The suitable techniques used in simulation DOE are those whose sampling points are spread over the entire design space. They are known as space filling design (Box and Hunter 1978), the two well-known methods are the Orthogonal arrays, and the Latin Hypercube design.

The appropriate sample size depends not only on the number of the parameter space but also on the computational time for a simulation run. This is due to the fact that a complex nonlinear function requires more sampling points. A proper way to use those DOE techniques in simulation is to maximize the minimum Euclidean distance between the sampling points so that the developer guarantees that the sampling points are spread along the complete regions in the parameter space (Jurecka 2007).

6.2.2 Interpolation

The process in which the deterministic discrete points are transformed into a connected continuous function is called interpolation. One important aspect for the Virtual Production Intelligence (VPI) systems is the availability of interpolating models that represent the process behavior (Reinhard 2013), which are the metamodels. In VPI, metamodeling techniques offer excellent possibilities for describing the process behavior of technical systems (Jurecka 2007; Chen 2001) since Meta-Modelling defines a procedure to analyze and simulate involved physical systems using fast mathematical models (Sacks 1989). These mathematical models create cheap numeric surrogates that describe cause-effect relationship between setting parameters as input and product quality variables as output for manufacturing processes. Among the available Meta-Modelling techniques are the Artificial Neural Networks (Haykin 2009), Linear Regression Taylor Expansion (Montgomery et al. 2012) Kriging (Jones 1998; Sacks 1989; Lophaven 1989), and the radial basis functions network (RBFNs). RBFN is well known for its accuracy and its ability to generate multidimensional interpolations for complex nonlinear problems (Rippa 1999; Mongillo 2010; Orr 1996). A Radial Basis Function Interpolation represented in Fig. 6.1 below is similar to a three layer feed forward neural network. It consists of an input layer which is modeled as a vector of real numbers, a hidden layer that contains nonlinear basis functions, and an output layer which is a scalar function of the input vector.

The output of the network f(x) is given by:

$$f(x) = \sum_{i=1}^{n} w_i h_i(x) \tag{6.1}$$

where n, h_i, w_i correspond to number of sampling points of the training set, the ith basis function, and the ith weight respectively. The RBF methodology was introduced in 1971 by Rolland Hardy who originally presented the method for the multiquadric (MQ) radial function (Hardy 1971). The method emerged from a cartography problem, where a bivariate interpolates of sparse and scattered data was needed to represent topography and produce contours. However, none of the existing interpolation methods (Fourier, polynomial, bivariate splines) were satisfactory because they were either too smooth or too oscillatory (Hardy 1990). Furthermore, the non-singularity of their interpolation matrices was not guaranteed. In fact, Haar's theorem states that the existence of distinct nodes for which the interpolation matrix associated with node-independent basis functions is singular in

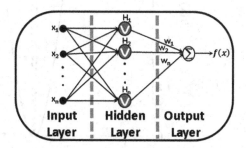

Fig. 6.1 Architecture of Radial Basis Function Network (RBFN). Solve for the weights w_i Given the parameter values x_i, the base functions $h_i(x)$ and the scalar output f(x) the

two or higher dimensions (McLeod 1998). In 1982, Richard Franke popularized the MQ method with his report on 32 of the most commonly used interpolation methods (Franke 1982). Franke also conjectured the unconditional non singularity of the interpolation matrix associated with the multi-quadric radial function, which was later proved by Micchelli (Micchelli 1986). The multi-quadric function is used for the basis functions h_i:

$$h_i(x) = \sqrt{1 + \frac{(x - x_i)^T (x - x_i)}{r^2}} \qquad (6.2)$$

where x_i and r represent the ith sampling point and the width of the basis function respectively. The shape parameter r controls the width of the basis function, the larger or smaller the parameter changes, the narrower or wider the function gets. This is illustrated in Fig. 6.2 below.

The learning of the network is performed by applying the method of least squares with the aim of minimizing the sum squared error with respect to the weights w_i of the model (Orr 1996). Thus, the learning/training is done by minimizing the cost function

$$C = \sum_{i=1}^{n} (y_i - f(x_i))^2 + \sum_{i=1}^{n} \lambda \cdot w_i^2 \rightarrow \min \qquad (6.3)$$

where λ is the usual regularization parameter and y_i are the criterion values at points i. Solving the equation above

Fig. 6.2 Multi-quadric function centered at $x_i = 0$ with different widths r = 0.1, 1, 2, 8

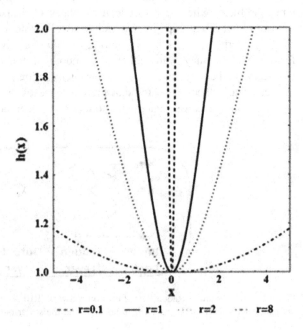

$$w = \left(H^T H + \Lambda\right)^{-1} H^T y \tag{6.4}$$

with

$$H = \begin{bmatrix} h_1(x_1) & h_2(x_1) & \cdots & h_n(x_1) \\ h_1(x_2) & h_2(x_2) & \cdots & h_n(x_2) \\ \vdots & \vdots & \ddots & \vdots \\ h_1(x_n) & h_2(x_n) & \cdots & h_n(x_n) \end{bmatrix} \qquad \Lambda = \begin{bmatrix} \lambda & 0 & \cdots & 0 \\ 0 & \lambda & \cdots & 0 \\ \vdots & \vdots & \ddots & \vdots \\ 0 & 0 & \cdots & \lambda \end{bmatrix} \tag{6.5}$$

and

$$y = (y_1, y_2, \ldots, y_n) \tag{6.6}$$

The chosen width of the radial basis function plays an important role in getting a good approximation. The following selection of the r value was proposed by Hardy (1971) and taken over for this study:

$$r = 0.81 \cdot d, \quad d = \frac{1}{n} \sum_{i=1}^{n} d_i \tag{6.7}$$

and d_i is the distance between the ith data point and its nearest neighbor.

6.2.3 Exploration

Visualization is very important for analyzing huge sets of data. This allows an efficient decision making. Therefore, multi-dimensional exploration or visualization tools are needed. 2D Contour plots or 3D cube plots can be easily generated by any conventional mathematical software. However nowadays, visualization of high dimensional simulation data remains a core field of interest. An innovative method was developed by Gebhardt (2013). In the second phase of the Cluster of Excellence "Integrative Production Technology for High-Wage Countries" the Virtual Production Intelligence (VPI). It relies on a hyperslice-based visualization approach that uses hyperslices in combination with direct volume rendering. The tool not only allows to visualize the metamodel with the training points and the gradient trajectory, but also assures a fast navigation that helps in extracting rules from the metamodel; hence, offering an user-interface. The tool was developed in a virtual reality platform of RWTH Aachen that is known as the aixCAVE. Another interesting method called the Morse-Smale complex can also be used. It captures the behavior of the gradient of a scalar function on a high dimensional manifold (Gerber 2010) and thus can give a quick overview of high dimensional relationships.

6.3 Applications

In this section, the metamodeling techniques are applied to different laser manufacturing processes. The first two applications (Laser metal sheet cutting and Laser
epoxy cut) where considered a data driven metamodeling process where models
where considered as a black box and a learning process was applied directly on the
data. The last two applications (Drilling and Glass Ablation) a model driven
metamodeiling process was applied.

The goal of this section is to highlight the importance of using the proper
metamodeling technique in order to generate a specific metamodel for every process. The developer should realize that generating a metamodel is a user demanding
procedure that involves compromises between many criteria and the metamodel
with the greatest accuracy is not necessarily the best choice for a metamodel. The
proper metamodel is the one which fits perfectly to the developer needs. The needs
have to be prioritized according to some characteristics or criteria which was
defined by Franke (1982). The major criteria are accuracy, speed, storage, visual
aspects, sensitivity to parameters and ease of implementation.

6.3.1 Sheet Metal Cutting with Laser Radiation

The major quality criterion in laser cutting applications is the formation of adherent
dross and ripple structures on the cutting kerf surface accompanied by a set of
properties like gas consumption, robustness with respect to the most sensitive
parameters, nozzle standoff distance and others. The ripples measured by the cut
surface roughness are generated by the fluctuations of the melt flow during the
process. One of the main research demands is to choose parameter settings for the
beam-shaping optics that minimize the ripple height and changes of the ripple
structure on the cut surface. A simulation tool called QuCut reveals the occurrence
of ripple formation at the cutting front and defines a measure for the roughness on
the cutting kerf surface. QuCut is developed at Fraunhofer ILT and the department
Nonlinear Dynamics of Laser Processing (NLD) at RWTH Aachen as a numerical
simulation tool for CW laser cutting taking into account spatially distributed laser
radiation. The goal of this use case was to find the optimal parameters of certain
laser optics that result in a minimal ripple structure (i.e. roughness). The 5 design
parameters of a laser optic (i.e. the dimensions of vector in formulas (6.1–6.5))
investigated here are the beam quality, the astigmatism, the focal position, and the
beam radius in x and y directions of the elliptical laser beam under consideration.
The properties of the fractional factorial design are listed in Table 6.1.

The selected criteria (i.e. y-vector in formulas (6.3–6.5)) was the surface
roughness (Rz in μm) simulated at a 7 mm depth of an 8 mm workpiece. The full
data set was 24948 samples in total. In order to assess the quality of the mathematical interpolation, 5 different RBFN metamodels were generated according to 5

Table 6.1 Process design domain

Beam Parameters	Minimum	Maximum	Sampling Points
Beam Quality M2	7	13	7
Astigmatism [mm]	15	25	9
Focal position [mm]	−8	20	11
Beam Radius x-direction [μm]	80	200	6
Beam Radius y-direction [μm]	80	200	6

randomly selected sample sets of size 1100, 3300, 5500, 11100 and 24948 data points from the total dataset. As shown in Fig. 6.3, the metamodels are denoted by Metamodel (A–E). Metamodel F, which is a 2D metamodel with a finer sampling points denoted by the blue points, is used as a reference for comparison.

A 2fold cross-validation method was then used to give the quality of meta-modeling, where 10 % of the training point sample was left out randomly of the interpolation step and used for validation purposes. The Mean Absolute Error (MAE) of the criterion surface roughness and the coefficient of determination (R2) were then calculated and compared to each other. The results are listed in Table 6.2 below.

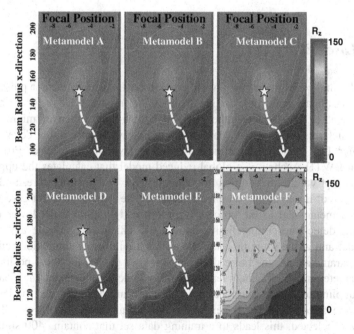

Fig. 6.3 2D Contour Plots of different metamodels at M2 = 10, Astigmatism = 25 mm, Beam Radius y = 134 μm. The polynomial linear regression metamodel (F) on the right contains more sampling points and is shown here for evaluation of the metamodel quality (A–E)

Metamodel	MAE/[μm]	R^2
A	7	13
B	15	25
C	−8	20
D	80	200
E	80	200

Table 6.2 Data of the metamodels

The results show that the quality of the metamodel is dependent on the number of sampling points; the quality is improved when the number of training points is increased. As visualization technique contour plots were used, which in their entirety form the process map. The star-shaped marker, denoting the seed point of the investigation, represents the current cutting parameter settings and the arrow trajectory shows how an improvement in the cut quality is achieved. The results show that in order to minimize the cutting surface roughness in the vicinity of the seed point, the beam radius in the feed direction x should be decreased and the focal position should be increased Eppelt and Al Khawli (2014) In the special application case studied here the minimum number of sampling points with an RBFN model is already a good choice for giving an optimized working point for the laser cutting process. These metamodels have different accuracy values, but having an overview of the generated tendency can support the developer with his decision making step.

6.3.2 Laser Epoxy Cut

One of the challenges in cutting glass fiber reinforced plastic by using a pulsed laser beam is to estimate achievable cutting qualities. An important factor for the process improvement is first to detect the physical cutting limits then to minimize the damage thickness of the epoxy-glass material. EpoxyCut, also a tool developed at Fraunhofer ILT and the department Nonlinear Dynamics of Laser Processing (NLD) at RWTH Aachen, is a global reduced model that calculates the upper and lower cutting width, in addition to other criteria like melting threshold, time required to cut through, and damage thickness. The goal of this test case was to generate a metamodel to the process in order to: (i) minimize the lower cutting width; (ii) detect the cutting limits; and (iii) to efficiently generate an accurate metamodel and at the same time use the minimal number of simulation runs. The process parameters are the pulse duration, the laser power, the focal position, the beam diameter and the Rayleigh length. In order to better understand the idea of the smart sampling technique, the focal position, the beam diameter, and the Rayleigh length were fixed. In order to generate a fine metamodel, a 20 level full factor design was selected, this leads to a training data set that contains 400 simulation runs in total illustrated as small white round points in Fig. 6.4.

Fig. 6.4 EpoxyCut process
map generated on a 20 full
factorial grid

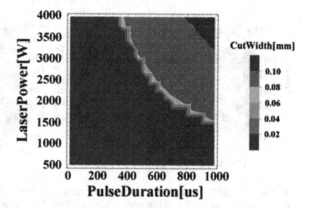

The metamodel takes the discrete training data set as an input, and provides the operator with a continuous relationship of the pulse duration and laser power (parameters) and cutting width (quality) as an output.

In order to address the first goal which is to minimize the lower cutting width, the metamodel above allows a general overview of the the 2D process model.

It can be clearly seen that one should either decrease the pulse duration or the laser power so that the cutting limits (the blue region determines the no cut region) are not achieved. The second goal was to include the cutting limits in the Meta-modelling generation. When performing a global interpolation techniques, the mathematical value that represents the no cutting regions (the user set it to 0 in this case) affects the global interpolation techniques.

To demonstrate this, a Latin Hypercube Design with 49 training points was used (big white circles) with an RBFN interpolation. The results are shown in Fig. 6.5 below.

From the results in Fig. 6.5, the developer can totally realize that process that contains discontinuities, or feasible and non-feasible points (in this case cut and no cut) should be classified first into feasible metamodel and a dichotomy metamodel.

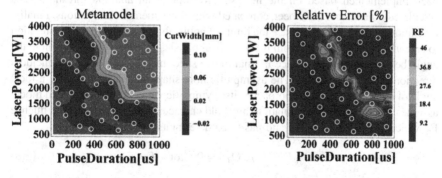

Fig. 6.5 An RBFN metamodel with Latin Hypercube design (49 sampling points) on the *left*, is compared to the EpoxyCut with a Relative Error in % shown to the *right*

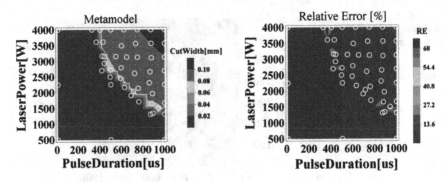

Fig. 6.6 An RBFN metamodel with smart sampling algorithm (also 49 sampling points) on the *left*, is compared to the EpoxyCut with a Relative Error in % shown to the *right*

The feasible metamodel improves the prediction accuracy in the cut region and the dichotomy metamodel states whether the prediction lies in a feasible or a non-feasible domain. This is one of the development fields that the authors of this paper are focusing on.

To address the third goal which is to efficiently generate an accurate metamodel and while using the minimal number of simulation runs. A smart sampling method at the department Nonlinear Dynamics of Laser Processing NLD at RWTH is currently being developed and will be soon published. The method is based on a classification technique with a sequential approximation optimization where training points are being iteratively sampled based on defined statistical measures. The results are shown in Fig. 6.6 below.

6.3.3 Sheet Metal Drilling

As example for heuristic approaches a reduced model for sheet metal drilling has been implemented based on the heuristic concept of an ablation threshold. The calculated hole shapes have been compared with experimental observations. Finally, by exploring the parameter space the limits of applicability are found and the relation to an earlier model derived from mathematical physics is revealed. Let Θ denote the angle between the local surface normal of the sheet metal surface and the incident direction of the laser beam. The asymptotic hole shape is characterized by a local angle of incidence Θ which approaches its asymptotic value Θ_{Th}. The reduced model assumes that there exists an ablation threshold characterized by the threshold fluence F_{th} which is material specific and has to be determined to apply the model.

$$\cos \Theta_{Th} F = F_{TH}, \Theta_{Th} = 0 \quad \text{for} \quad F < F_{TH} \tag{6.8}$$

One single simulation run is used to estimate the threshold fluence F_{Th} where the width of the drill at the bottom is fitted. As consequence the whole asymptotic shape of the drilled hole is calculated and is illustrated in Fig. 6.7 below.

Finally, classification of sheet metal drilling can be performed by identification of the parameter region where the drill hole achieves its asymptotic shape. It is worth to mention, that for the limiting case of large fluence $F \gg F_{th}$ the reduced model is well known from literature (Schulz 1986, 1987) and takes the explicit form:

$$\frac{dz(x)}{dx} = \frac{F}{F_{TH}}, z(x_{Th}) = 0, F \gg F_{Th} \tag{6.9}$$

where z(x) are is the depth of drilled wall and x is the lateral coordinate with respect to the laser beam axis.

6.3.4 Ablation of Glass

As an example for classification of a parameter space we consider laser ablation of glass with ultrashort pulses as a promising solution for cutting thin glass sheets in display industries. A numerical model describes laser ablation and laser damage in glass based on beam propagation and nonlinear absorption as well as generation of free electrons (Sun 2013). Free-electron density increases until reaching the critical electron density $\rho_{crit} = (\omega^2 m_e \varepsilon_0)/e = 3.95 \times 1021\,cm^{-3}$, which yields the ablation threshold.

The material near the ablated-crater wall will be modified due to the energy released by high-density free-electrons. The threshold electron density ρ_{damage} for laser damage is a material dependent quantity, that typically has the value $\rho_{damage} = 0.025\,\rho_{crit}$ and is used as the damage criterion in the model. Classification of the parameter region where damage and ablation takes place reveal the threshold in terms of intensity, as shown in Fig. 6.8 below, change from an intensity threshold at ns-pulses to a fluence threshold at ps-pulses.

Fig. 6.7 Cross section and shape (*solid curve*) of the drill hole calculated by the reduced model for sheet metal drilling

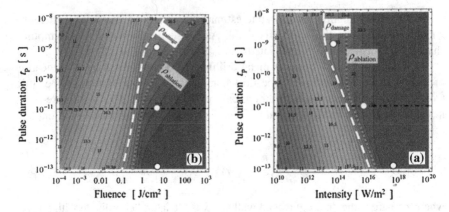

Fig. 6.8 Cross section and shape (*solid curve*) of the drill hole calculated by the reduced model for sheet metal drilling

6.4 Conclusion and Outlook

This contribution is focused on the application of the Meta-Modeling techniques towards Virtual Production Intelligence. The concept of Meta-modelling is applied to laser processing, e.g. sheet metal cutting, sheet metal drilling, glass cutting, and cutting glass fiber reinforced plastic. The goal is to convince the simulation ana-lysts to use the metamodeling techniques in order to generate such process maps that support their decision making. The techniques can be applied to almost any economical, ecological, or technical process, where the process itself is described by a reduced model. Such a reduced model is the object of Meta-Modeling and can be seen as a data generating black box which operates fast and frugal. Once an initial reduced model is set then data manipulation is used to evaluate and to im-prove the reduced model until the desired model quality is achieved by iteration. Hence, one aim of Meta-Modeling is to provide a concept and tools which guide and facilitate the design of a reduced model with the desired model quality. Evalu-ation of the reduced model is carried out by comparison with rare and expensive data from more comprehensive numerical simulation and experimental evidence. Finally, a Meta-Model serves as a user friendly look-up table for the criteria with a large extent of a continuous support in parameter space enabling fast exploration and optimization.

The concept of Meta-Modeling plays an important role in improving the quality of the process since: (i) it allows a fast prediction tool of new parameter settings, providing mathematical methods to carry out involved tasks like global optimization, sensitivity analysis, parameter reduction, etc.; (ii) allows a fast user interface exploration where the tendencies or patterns are visualized supporting intuition; (iii) replaces the discrete data of current conventional technology tables or catalogues which are delivered almost with all the manufacturing machines for a good operation by continuous maps.

It turns out that a reduced model or even a Meta-Model with the greatest accuracy is not necessarily the "best" Meta-Model, since choosing a Meta-Model is a decision making procedure that involves compromises between many criteria (speed, accuracy, visualization, complexity, storage, etc.) of the Meta-Model quality. In the special application case studied here the minimum number of sampling points with a linear regression model is already a good choice for giving an optimized working point for sheet metal cutting, if speed, storage and fast visualization are of dominant interest. On the other hand when dealing with high accuracy goals especially when detecting physical limits, smart sampling techniques, nonlinear interpolation models and more complex metamodels (e.g. with classification techniques) are suitable.

Further progress will focus on improving the performance of generating the metamodel especially developing the smart sampling algorithm, and verifying it on other industrial applications. Additional progress will focus on allowing the creation of a metamodel that handles distributed quantities and not only scalar quantities. Last but not least, these metamodels will be interfaced to global sensitivity analysis technique that helps to extract knowledge or rules from data.

Acknowledgments The investigations are partly supported by the German Research Association (DFG) within the Cluster of Excellence "Integrative Production Technology for High-Wage Countries" at RWTH Aachen University as well as by the European Commission within the EU-FP7-Framework (project HALO, see http://www.halo-project.eu/).

References

Asimov D (2011) The grand tour: a tool for viewing multidimensional data, SIAM J. Sci. Stat. Comput. 1985;6(1):128–143

Auerbach T, M. Beckers, G. Buchholz, U. Eppelt,Y. Gloy, P. Fritz, T. Al Khawli, S. Kratz, J. Lose, T. Molitor, A. Reßmann, U. Thombansen, D. Veselovac, K. Willms, T. Gries, W. Michaeli, C. Hopmann, U., R.Schmitt, and F Klocke (2011) Meta-modeling for Manufacturing, ICIRA 2011, Part II, LNAI 7102, pp. 199–209, 2011.

Belkin M, Niyogi P (2003) Laplacian eigenmaps for dimensionality reduction and data representation, Neural Comput;15(6):1373–1396.

Bellman R (1957) Dynamic programming. Number ISBN 978-0-691-07951-6. Princeton University Press

Box P, Hunter G (1978). Statistics for Experimenters: An Introduction to Design, Data Analysis, and Model Building. John Wiley and Sons

Buckingham E (1914) On physically similar systems; illustrations of the use of dimensional equations, Physical Review 4, 345–376

Chen W, Jin R, Simpson T (2001) Comparative Studies of Metamodelling Techniques under Multiple Modeling Criteria

Currie N (2005) Conceptual Design: Building a Social Conscience, AIGA, November 1

Eppelt U, Al Khawli T (2014) Metamodeling of Laser Cutting, Presentation and Proceedings paper In: ICNAAM—12th International Conference of Numerical Analysis and Applied Mathematics, September 22–28, 2014, Rodos Palace Hotel, Rhodes, Greece, Preprint Fraunhofer ILT

Ferrari D, Borrotti M (2014), Response improvement in complex experiments by coinformation composite likelihood optimization. Statistics and Computing, Volume 24, Issue 3, pp 351–363

Franke R (1982), Smooth interpolation of scattered data by local thin plate splines, Comp. & Maths. with Appls. 8, 273

Gebhardt S, Al Khawli T, Hentschel B, Kuhlen T, Schulz W. (2013) Hyperslice Visualization of Metamodels for Manufacturing Processes, IEEE Visualization Conference (VIS): Atlanta, GA, USA, 13 Oct—18 Oct 2013

Gerber S, Bremer T, Pascucci V, Whitaker R (2010) Visual Exploration of High Dimensional Scalar Functions, IEEE Trans Vis Comput Graph. 16(6): 1271–1280

Hardy R (1990) Theory and applications of the multiquadric biharmonic method: 20 years of discovery 1968

Hardy R (1971) Multiquadric equations of topography and other irregular surfaces. Journal of Geophysical Research, 76(8):1905–1915

Haykin S (2009) S. Neural Networks and Learning Machines (3rd Edition), Prentice Hall

Jones D, Schonlau M, Welch W, (1998) Efficient Global Optimization of Expensive Black-Box Functions. Journal of Global Optimization Vol. 13, 455–492

Jurecka F. (2007) Robust Design Optimization Based on Metamodeling Techniques, Shaker Verlag

Kleijnen J P, Sanchez S M, Lucas T W, Cioppa T M (2005): State of the art review: A user's guide to the brave new world of designing simulation experiments. INFORMS Journal on Computing 17(3), 263–289

Lophaven S, Nielsen H B, Søndergaard J (2002) Dace a MATLAB Kriging Toolbox Version 2.0

Micchelli C A (1986) Interpolation of scattered data: distance matrices and conditionally positive definite functions, Constr. Approx.2, 11

McLeod G (1998). Linking Business Object Analysis to a Model View Controller Based Design Architecture, Proceedings of the Third CAiSE/IFIP 8.1 International Workshop on Evaluation of Modeling Methods in Systems Analysis and Design EMMSAD'98, Pisa, Italy.

Mongillo M (2010) Choosing basis functions and shape parameters for radial basis function methods, Comput. Math. Appl. 24, pp. 99–120

Montgomery D C, Peck E,Vining G (2012). Introduction to linear regression analysis (Vol. 821). Wiley.

Orr M (1996): Introduction to radial basis function networks

Reinhard R, Al Khawli T, Eppelt U, Meisen T, Schilberg D, Schulz W, Jeschke S (2013) How Virtual Production Intelligence Can Improve Laser-Cutting Planning Processes, In: ICPR 22—Systems Modeling and Simulation (p. 122)

Reinhard R, Al Khawli T, Eppelt U, Meisen T, Schilberg D, Schulz W, Jeschke S (2014) The Contribution of Virtual Production Intelligence to Laser Cutting Planning Processes, In: Enabling Manufacturing Competitiveness and Economic Sustainability (pp. 117–123). Springer International Publishing. 2014

Rippa S (1999): An algorithm for selecting a good value for the parameter c in radial basis function interpolation, Adv Comput Math 11(2–3), 193–210

Sacks J, Welch W J, Mitchell T J, Wynn H (1989) Design and Analysis of Computer Experiments, Statistical Science, Vol. 4, No. 4, pp. 409–423

Schulz W, Simon G, Vicanek M (1986) Ablation of opaque surfaces due to laser radiation, J.Phys. D: Appl. Phys. 19 173–177

Schulz W, Simon G, Urbassek H, Decker I (1987) On laser fusion cutting of metals, J.Phys.D: Appl.Phys. 20 481–488

Schwarz G A (1978) Estimating the dimension of a mode. Statist. 1978;6(2):461–464

Sun M, Eppelt U, Russ S, Hartmann C, Siebert C, Zhu J, Schulz W (2013) Numerical analysis of laser ablation and damage in glass with multiple picosecond laser pulses. Optics express, 21(7), pp. 7858–7867

Chapter 7
Designing New Forging Steels by ICMPE

Wolfgang Bleck, Ulrich Prahl, Gerhard Hirt and Markus Bambach

Abstract Any production is based on materials. Material properties are of utmost importance, both for productivity as well as for application and reliability of the final product. A sound prediction of materials properties thus is highly important. For metallic materials, such a prediction requires tracking of microstructure and properties evolution along the entire component process chain. In almost all nature and engineering scientific disciplines the computer simulation reaches the status of an individual scientific method. Material science and engineering joins this trend, which permits computational material and process design increasingly. The Integrative Computational Materials and Process Engineering (ICMPE) approach combines multiscale modelling and through process simulation in one comprehensive concept. This paper addresses the knowledge driven design of materials and processes for forgings. The establishment of a virtual platform for materials processing comprises an integrative numerical description of processes and of the microstructure evolution along the entire production chain. Furthermore, the development of ab initio methods promises predictability of properties based on fundamentals of chemistry and crystallography. Microalloying and Nanostructuring by low temperature phase transformation have been successfully applied for various forging steels in order to improve component performance or to ease processing. Microalloying and Nanostructuring contribute to cost savings due to optimized or substituted heat treatments, tailor the balance of strength and toughness or improve the cyclic. A new materials design approach is to provide damage tolerant matrices and by this to increase the service lifetime. This paper deals with the numerically based design of new forging steels by microstructure refinement, precipitation control and optimized processing routes.

W. Bleck (✉) · U. Prahl
Department of Ferrous Metallurgy (IEHK), RWTH Aachen University,
Intzestr. 1, 52072 Aachen, Germany
e-mail: bleck@iehk.rwth-aachen.de

G. Hirt · M. Bambach
Metal Forming Institute (IBF), RWTH Aachen University, Intzestr. 10,
52056 Aachen, Germany
e-mail: hirt@ibf.rwth-aachen.de

C. Brecher (ed.), *Advances in Production Technology*,
Lecture Notes in Production Engineering, DOI 10.1007/978-3-319-12304-2_7

7.1 Introduction

For many applications, the material carries the properties and therefore is of vital importance for the product usability and also for the further innovation potential (acatech 2008). The development of new steels with improved properties or new production processes with ecologically and economically optimized process chains is a high priority for ensuring quality of life and competitiveness (ICME-NRC 2008).

In research and development of simulation methods are used increasingly. This trend is based on both the development of models and methods as well as on the increase of available computing resources. Recent developments allow for the implementation of computationally and memory intensive simulations for complex physical-chemical phenomena in multicomponent systems and real structures. By this, new simulation methods offer a relevant reduction of development time, support the sustainable use of resources (raw materials, energy, time), and help to avoid mistakes (Schuh et al. 2007).

Figure 7.1 shows for different situations of material development the opposite dependence of degree of novelty with the associated risks, costs and times on the one hand and the level of familiarity on the other hand (Moeller 2008). For current steel development examples, the modelling approaches used and the approximate beginning of industrial implementation and modelling are given. It is shown that the modelling is increasingly early integrated in the industrial development process.

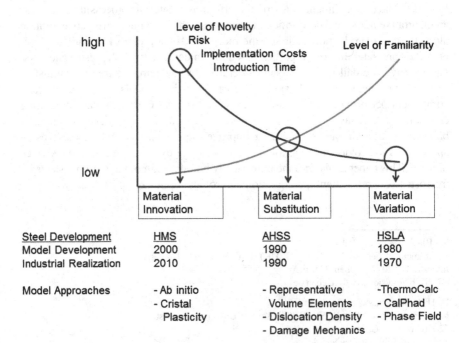

Fig. 7.1 Decision situation and design methods for recent developments of steels

This trend correlates with the development of a descriptive modelling using thermodynamic databases for the identification of material variations (e.g. microalloyed steels—HSLA) towards a more predictive simulation using ab initio methods and crystal plasticity. These predictive methods are now available in a way that they can be used to develop new classes of materials such as high manganese TWIP steels (HMS). The RWTH takes in the Collaborative Research Center (SFB) 761 "Steel—ab initio" in cooperation with the MPIE in Dusseldorf active part in the combined development of modelling methods and materials (v. Appen et al. 2009).

7.2 Interplay of Various Modelling Approaches

Through the development of models and methods in materials science and engineering new insight knowledge and new design ideas for the complex material system steel alloy are generated. However, for the various models at different scales along the process chain an inter-communicating approach is needed. At RWTH Aachen University the project AixViPMaP® was started with the goal to design a modular, standardized, open and extensible simulation platform offering a focusable, integrative simulation of process chains for material production, treatment and deployment (Steinbach 2009). Figure 7.2 illustrates the concept of the platform project to implement coordinated communication between the different scales and processes

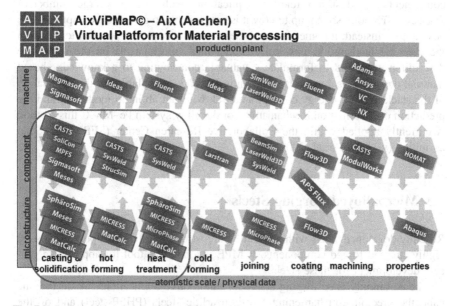

Fig. 7.2 Layout of the virtual platform for materials processing. Indicated by a frame: simulation of the process chain "gear component"—focusing only on the most relevant production steps (Schmitz and Prahl 2012)

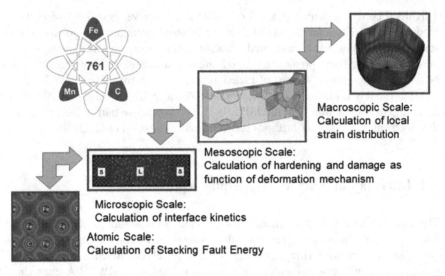

Fig. 7.3 "Scale-hopping" approach to a knowledge-based materials development in the SFB 761 "Steel—ab initio" (v. Appen et al. 2009)

along the production chain of a gear, where only the relevant processes (here surrounded with circles) by the appropriate simulation tools (dark outline) are modelled.

However, still the number of considered scales, process steps and chemical components in industrially relevant applications leads to a very large number of degrees of freedom, so that up to now it is not possible to generate a comprehensive description. Instead, it comes in the sense of "scale-hopping" approach to focus on the core mechanisms, to physically model these mechanisms on the respective description scale and to formulate a valuable contribution to a knowledge-based material design. This approach is shown in Fig. 7.3 exemplarily for the application of the stacking fault energy concept as a link between ab initio modelling and the prediction of deformation mechanisms. For the alloy system Fe–Mn–C this method is currently applied within the Collaborative Research Center (SFB) 761 "Steel ab initio" (v. Appen et al. 2009).

7.3 Microalloyed Forging Steels

Recently developed steels with a bainitic microstructure offer great possibilities for highly stressed forged components. ICMPE is a decisive tool for appropriate process development for these steels.

The commonly used forging steels for automotive applications are on the one hand the precipitation hardening ferritic-pearlitic steels (PHFP-steel) and on the other hand the quenched and tempered (Q&T) forging steels. The advantages of these PHFP steels compared to Q&T steels are the elimination of an additional heat

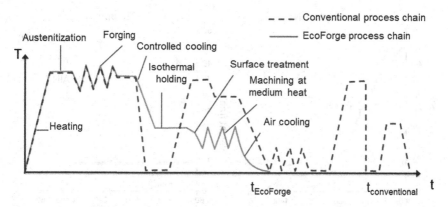

Fig. 7.4 Time-temperature sequence for conventional Q + T forging steels (*red*) and for bainitic forging steels (*green*)

treatment step which includes a hardening, tempering and stress relieving due to a controlled cooling directly after hot forging (Fig. 7.4) and an improved machin-ability (Langeborg et al. 1987; Gladman 1997). However, forging steels with fer-ritic/pearlitic microstructures show inferior values of yield strength and toughness compared to the Q&T steels.

In order to improve the toughness while maintaining high strength values a bainitic microstructure can be employed (Honeycombe and Bhadeshia 1995; Bhadeshia 2001; Wang et al. 2000). Figure 7.5 shows the achievable tensile strengths in dependence of the microstructure for PHFP-M and high strength ductile bainitic (HDB) steels. The different microstructures are mainly adjusted by

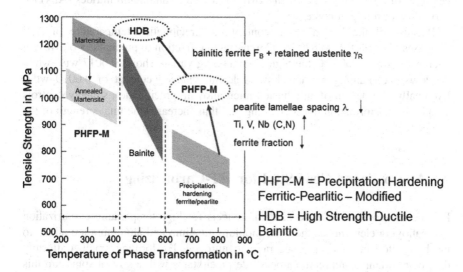

Fig. 7.5 Tensile strength values in dependence of microstructure for different forging steels

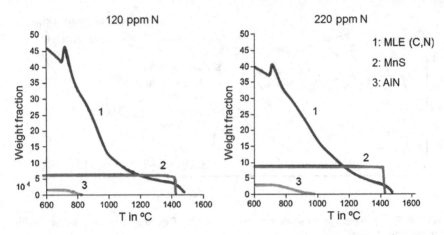

Fig. 7.6 Simulated fraction of precipitates in microalloyed AFP steel for varying nitrogen contents (Erisir et al. 2008)

choosing the right temperature for the phase transformation of the supercooled austenite.

The increase in strength for the PHFP-M steel is achieved by reduction of the ferritic volume fraction, the decrease in the pearlite lamellae spacing λ and the addition of the microalloying elements Nb and Ti which results in additional precipitates besides the vanadium nitrides (Langeborg et al. 1987; Bleck et al. 2010). For the design of these steels thermodynamic modelling utilizing the ThermoCalc software (Andersson et al. 2002) offers a crucial contribution to adjust the optimal microalloying and nitrogen composition. Figure 7.6 shows the precipitation temperatures of microalloying elements (MLE) as well as aluminum nitrides (AlN) for two different nitrogen contents.

Because of the low nitrogen content the precipitation temperature of AlN decreases from 980 °C to 820 °C. Comparing the fraction of precipitates of MLE at a temperature of 1,000 °C the high N containing variant shows 0.0017 wt% while the low N containing variant shows relevant reduced content of 0.0010 wt%. Eventually, the design of an adjusted microalloying precipitation strategy controls the phase transformation during cooling and thus increases the final strength of the component.

7.4 Microalloyed Gear Steel for HT-Carburizing

For the development of case hardening steels for high-temperature carburization microalloying elements as there are niobium, titanium and aluminum are added to the base alloy in an appropriate ratio to nitrogen. By forming small, uniformly distributed titanium-niobium carbonitride precipitates with a size of some nm this concept offers to ensure the stability of the austenite grain size for carburizing

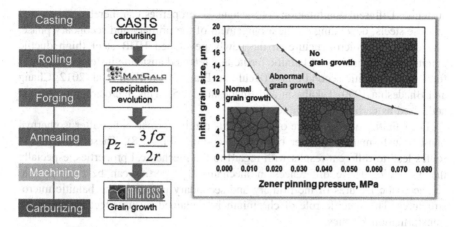

Fig. 7.7 Precipitation management and process window identification for microalloyed steel for high-temperature carburization (Prahl et al. 2008)

temperatures higher than 1,000 °C. The austenite grain size is decisive for the cyclic properties of the final component; therefore inhomogeneous grain growth has to be avoided. Consequently, the precipitation behaviour has to be controlled along the entire process chain from the steel shop via casting, forming, heat treatments to the manufacturing of the gear component.

For this example, thermodynamic modelling provides the key for the design process of material and process chain (Fig. 7.7). Here, the program MatCalc is utilized allowing to follow the precipitation evolution along the production chain continuous casting, rolling, forging, annealing, and final carburizing and thus to control the grain size evolution by grain boundary pinning (Kozeschnik et al. 2007).

Figure 7.7 shows the principal design concept; that is to identify a process window for the high-temperature carburization utilizing different simulation programs within a multiscale approach. In this example regions of different grain size stability are calculated as a function of Zener pinning pressure and initial austenite grain size for a thermal treatment of 1 h carburization at 1,050 °C. In this calculation the chemical composition and the precipitation state determines the Zener pinning pressure, which in turn is determined in a thermodynamic calculation (Prahl et al. 2008).

7.5 Bainitic Steels

The variety of different bainitic morphologies requests for an aligned thermal treatment after forging in order to achieve the maximum performance in terms of mechanical properties. In dependence of the alloying concept and heat treatment bainite is composed of different microstructural components like the ferritic primary phase and the secondary phase, which consists of either carbides, martensite and/or austenite. Different combinations of mechanical properties can thereby be adjusted in these steels, depending on the arrangement of the primary and secondary phase. The aimed for microstructure in the newly developed HDB steel (high ductile bainite) consists mainly of bainitic ferrite and retained austenite instead of carbides form as the bainitic second phase (Keul and Blake 2011; Keul et al. 2012; Chang and Bhadeshia 1990; Takahashi and Bhadeshia 1995). This microstructure is often addressed as carbide free bainite.

The bainitic microstructure of these steels can be formed either after isothermal phase transformation or after continuous cooling (Fig. 7.8). These two process routes lead to different results with regard to the mechanical properties, especially the Y/T-ratio. These differences in mechanical properties can be correlated to characteristic features of the primary and secondary phases of the bainitic microstructures. The specific role of chromium is explained by its effect on the phase transformation kinetics.

The phase transformation kinetics and the microstructure evolution during bainite formation can be simulated by means of multi-phase field simulation approach (Steinbach 2009). Figure 7.9 displays the simulated and the experimental-observed carbide precipitation within the lower bainite microstructure formed at 260 °C in 100Cr6 steel. In this bearing steel, the nano-sized carbide precipitation within the lower bainite microstructure tends to adopt a single crystallographic variant in a bainitic ferrite plate and this is different from the carbide precipitation within the tempered martensitic microstructures, where multiple crystallographic variants are preferred.

Lower bainite forms in the lower temperature range of the bainite transformation field between 400 and 250 °C. Because of the low transformation temperature, carbon diffusion is strongly restricted, so that the carbon that is insoluble in ferrite cannot diffuse out of the ferrite plates. As a result, in lower bainite, the

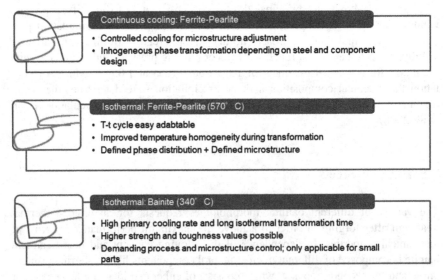

Continuous cooling: Ferrite-Pearlite

- Controlled cooling for microstructure adjustment
- Inhogeneous phase transformation depending on steel and component design

Isothermal: Ferrite-Pearlite (570° C)

- T-t cycle easy adabtable
- Improved temperature homogeneity during transformation
- Defined phase distribution + Defined microstructure

Isothermal: Bainite (340° C)

- High primary cooling rate and long isothermal transformation time
- Higher strength and toughness values possible
- Demanding process and microstructure control; only applicable for small parts

Fig. 7.8 Alternative cooling strategies for forging steels after deformation

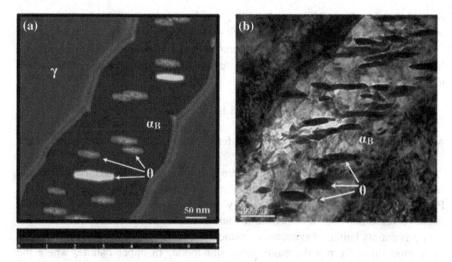

Fig. 7.9 Comparison of the simulated nano-sized carbide precipitation within lower bainite microstructure at 260 °C in 100Cr6 steel using multi-phase field approach with the experimental observation by TEM. **a** multi-phase field simulation **b** TEM bright field micrograph. The *colour bar* in (a) ranges from 0 wt% to 7 wt% (Song et al. 2013a, b)

diffusion-controlled sub-step of the transformation reaction consists of a precipitation of carbide particles within the growing ferrite plates. In doing so, carbides preferably assume an angle of approximately 60° from the ferrite axis. This angle is a result of the preferred nucleation on the intersection between the (101)-shear planes of ferrite with the bainite/austenite phase boundary.

In lower bainite the C precipitation does not necessarily lead to the equilibrium phase cementite, instead the more easily nucleated ε carbide may precipitate, or ε carbide precipitation precedes the formation of Fe_3C. The Atom Probe Tomography (APT) images in Fig. 7.10 (left) shows the 3D carbon atomic map and 1D

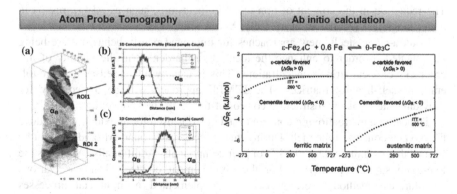

Fig. 7.10 Carbide precipitation within bainite in steel 100Cr6. *Left* atom probe results, indicating θ and ε carbides. *Right* ab initio based Gibbs energy calculation for precipitation in ferritic or austenitic matrix (Song et al. 2013a, b)

concentration profiles of lower bainite in 100Cr6 steel. It provides a local overview of the carbon distribution in bainitic ferrite matrix and carbides.

After long holding period, ε carbides transform into the equilibrium phase Fe_3C. In steels, the obvious reaction in an iron matrix is the transition between ε carbide/iron and cementite (θ),

$$\varepsilon\text{-Fe}_{2.4}\text{C} + 0.6\ \text{Fe} \rightleftarrows \theta\text{-Fe}_3\text{C},$$

where Fe is either bcc iron in a bainitic-ferritic matrix at low temperatures or fcc iron in austenite at higher temperatures.

Figure 7.10 (right) shows the Gibbs free reaction energies between $\varepsilon\text{-Fe}_{2.4}\text{C}$ and cementite $\theta\text{-Fe}_3\text{C}$ as a function of temperature in a ferritic and an austenitic matrix. Positive value of the Gibbs free energy indicates an ε favoured region and a negative value indicates a cementite favoured regime. In lower bainite, where the matrix is mainly bainitic ferrite, the formation of $\theta\text{-Fe}_3\text{C}$ and $\varepsilon\text{-Fe}_{2.4}\text{C}$ has nearly the same probability from a thermodynamic standpoint. In upper bainite, where the matrix is austenite, however, the formation of cementite is clearly preferred at any temperature. The theoretical calculations reveal that the formation of $\varepsilon\text{-Fe}_{2.4}\text{C}$ benefits from a ferritic matrix and thus ε carbide is more prone to precipitate from lower bainite than from upper bainite.

7.6 Al-Free Gear Steel

Materials development for improved strength-formability balances or higher toughness requirements must follow two major routes: either avoiding detrimental microstructural features and/or improving the matrix to enable a higher tolerance for local microstructural irregularities or degradations. In most applications, the plan is to avoid detrimental microstructural features by the improvement of the internal cleanliness, because inclusions are considered to be the main crack origin. The reduction in the content of non-metallic inclusions, such as Al_2O_3, results in better toughness (Melander et al. 1991; Murakami 2012).

In ultra-clean steels, new approaches for improved matrix behaviour are being investigated in order to enhance the local strain hardening in the vicinity of microcracks or local stress concentrations. This is usually addressed as damage-tolerant or self-healing matrices. The ICMPE approach will be a necessity for providing the right microstructure control of this new steel concept.

Typically, a microalloying concept based on Al is used for deoxidation to reduce the oxygen content in the melt. During this process hard, round Al-oxides might be formed that eventually limit the life of gear components. Additionally, Al affects the fine-grain stability positively. For the improvement of steel cleanness, various metallurgical methods were successfully implemented in industrial processes (Zhang and Thomas 2003). A material-based approach for the improvement of the steel cleanness can be achieved by reducing the Al content. This concept was successfully evaluated for bearing steels (Theiry et al. 1997). However, such low Al contents cannot ensure fine-grain stability in case hardening steels.

By using a combined thermodynamical and continuum mechanical multi-scale simulation approach a new alloying concept for steel 25MoCr4, alloyed with Nb and with reduced Al content has been developed (Konovalov et al. 2014). The aim of the investigation is to improve the oxide steel cleanness by reducing the Al content and in parallel increase the fine-grain stability at a high carburizing temperature of about 1,050 °C by substitution of Al by Nb. The development of an Al-free alloying concept is based on thermodynamical calculations to control the precipitation state in the relevant temperature range. Figure 7.11 shows the calculation of the maximum possible precipitation amount and its dependence on temperature carried out using the thermodynamic software Thermo-Calc.

For a first approximation, the calculation for the reduced Al content steel was performed at 30 ppm Al and compared with a reference material. The volume fraction of particles at the carburization temperature of 1,050 °C (T_A) is noticeably lower in comparison to the reference material. In the following calculations the Nb-content was increased step by step in order to achieve an equal volume fraction as compared to the reference material.

The simulation shows that the micro-alloying phases can be stable in the liquid-solid region and this can lead to the formation of coarse primary particles. Such coarse particles reduce cleanness and are not effective for fine grain stability. Thus, additional calculations were performed for a reduced Ti content of around 10 ppm. The target amounts of 800–900 ppm Nb, <30 ppm Al and approximately 10 ppm Ti has been determined. Finally, target area for the Al-free composition with the expected fine grain stability is shown as the hatched area in Fig. 7.11.

Fig. 7.11 Determination of target alloy system for Al-free carburizing steel by varying Nb- and Ti-contents (Konovalov et al. 2014)

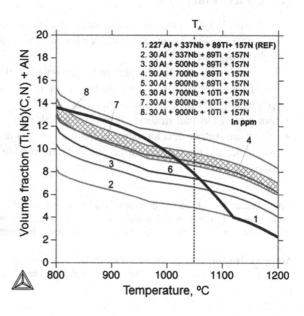

1. 227 Al + 337Nb + 89Ti + 157N (REF)
2. 30 Al + 337Nb + 89Ti + 157N
3. 30 Al + 500Nb + 89Ti + 157N
4. 30 Al + 700Nb + 89Ti + 157N
5. 30 Al + 900Nb + 89Ti + 157N
6. 30 Al + 700Nb + 10Ti + 157N
7. 30 Al + 800Nb + 10Ti + 157N
8. 30 Al + 900Nb + 10Ti + 157N
in ppm

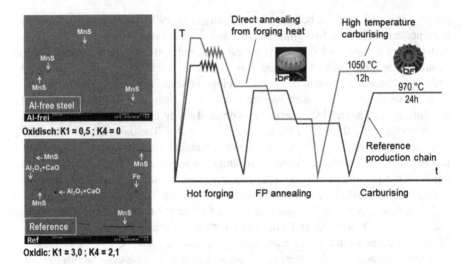

Fig. 7.12 Al-free gear steel for high temperature annealing yields improved cleanliness and shortens the production route by direct annealing from forging heat combined with short time carburizing (Konovalov et al. 2014)

For validation, a laboratory melt has been made and investigated regarding steel cleanness and fine-grain stability at high carburizing temperatures for different process routes (Fig. 7.12).

7.7 Conclusions

- A focused virtual description of process chains leads to a significant increase in planning quality, because knowledge-based predictions of material and process behaviour are possible.
- A modular, standardized, open and extensible simulation platform is a key to a significant increase planning efficiency in the development, production and processing of materials and components.
- For a truly "virtual material development" ab initio methods are essential.
- There are further developments in the field of 3-D dislocation dynamics needed to predict the mechanical properties and deformation of materials on a physical basis.

Acknowledgments The presented work is based on results that have been funded within various public projects. In detail the authors acknowledge the financial support within the following projects

- "Integrative Production Technologies in High Wage Countries" (DFG—Cluster of Excellence)
- "Steel ab initio" (DFG—Collaborative Research Center SFB 761)
- "New Steels and optimized Process Chain for high strength steels in forged structural components" (AVIF A 228)
- "Efficient process chains and new high strength (bainitic) steels for flexible production of highly loaded structural components" (IFG 260 ZN)
- "DiffBain" (ICAMS)
- "Al-free, Nb-stabilised Carburizing Steel for large Gears" (AVIF A 286).

References

acatech: Materialwissenschaft und Werkstofftechnik in Deutschland—Empfehlungen zu Profilierung, Lehre und Forschung; Fraunhofer IRB Verlag, Stuttgart (2008), ISBN 978-3-8167-7913-1

Andersson, J.O. et al.: Thermo-Calc, DICTRA, Computational Tools for materials Science; CalPhad 26 (2002), 273–312

Bhadeshia, H. K. D. H.: Bainite in steels—Transformations, microstructure and properties, 2nd edition, The Institute of Metals, 2001

Bleck, W.; Keul, C.; Zeislmair, B.: Entwicklung eines höherfesten mikrolegierten ausscheidungshärtenden ferritisch/perlitischen Schmiedestahls AFP-M, Schmiede-Journal (2010) März, S. 42–44

Chang, L. C.; Bhadeshia, H. K. D. H.: Mater. Sci. Tech., 1990, Vol. 6, pp. 592–603

Erisir, E.; Zeislmair, B; Keul, C.; Gerdemann, F.; Bleck, W.: "New developments for microalloyed high strength forging steels", 19th Int. Forging Congress, 2008, Chicago

Gladman, T.: The Physical Metallurgy of Microalloyed Steels, The Institute of Materials, London (1997), 341–348

Honeycombe, R. W. K.; Bhadeshia, H. K. D. H.: Steels Microstructure and Properties, 2nd edition, Edward Arnold, London, 1995

ICME-NRC—Committee on Integrated Computational Materials Engineering, National Research Council: Integrated Computational Materials Engineering: A Transformational Discipline for Improved Competitiveness and National Security; National Academic Press, Washington, D. C. (2008), ISBN: 0-309-12000-4

Keul, C.; Bleck, W.: New Microalloyed Steels for Forgings, 6th Int. Conf. on High Strength Low Alloy Steels (HSLA Steels'2011), 31.05.-02.06.2011, Beijing, China. Journal of Iron and Steel Research International 18 (2011) Supplement 1-1 (May 2011), S. 104–111

Keul, C.; Wirths, V.; Bleck, W.: New bainitic steels for forgings, Archives of Civil and Mechanical Engineering 12 (2012) Nr. 2, S. 119–125

Konovalov, S.; Sharma, M.; Prahl, U.: Nb Microsegregation and Redistribution during Casting and Forging in Al-free Case Hardening Steel, Proceedings of Int. Conf. on Rolling and Forging, 7.-9.5.2014, Milano, Italy

Kozeschnik, E. et al.: Computer Simulation of the Precipitate Evolution during Industrial Heat Treatment of Complex Alloys; Materials Science Forum 539–543 (2007), 2431–2436

Langeborg, R.; Sandberg, O.; Roberts, W.; in: G. Krauss and S. K. Banerji (eds.), Fundamentals of Microalloying Forging Steels, TMS, Warrendale, PA (1987), 39–54

Melander, A.; Rolfsson, M. et al.: Influence of inclusion contents on fatigue properties of SAE 52100 bearing steels, Scandinavian J. of Metallurgy 20 (1991), 229–244

Moeller, E.: Handbuch der Konstruktionswerkstoffe; Carl Hanser Verlag, München (2008), ISBN 978-3-446-40170-9

Murakami, Y.: Material defects as the basis of fatigue design, Int. J. of Fatigue 40 (2012), 2–10

Prahl, U.; Erisir, E.; Rudnizki, J.; Konovalov, S.; Bleck, W.: Mikrolegierte Einsatzstähle für die Hochtemperatur-Aufkohlung in Experiment und Simulation, 49. Arbeitstagung „Zahnrad- und Getriebeuntersuchungen" des WZL, 23.-24.04.08, Aachen, Germany

Schmitz, G. J.; Prahl, U.: Towards a virtual platform for materials pro-cessing, JOM 61 (2009) 5, 19

Schmitz, G.J.; Prahl, U. (eds.): Integrative Computational Materials Engineering—Concepts and Application of a Modular Simulation Platform, Wiley-VCH (2012), ISBN: 978-3-527-33081-2

Schuh, G.; Klocke, F.; Brecher, C.; Schmidt, R. (Hrsg.): Excellence in Production; Apprimus Verlag, Aachen (2007), ISBN 978-3-940565-00-6

Song, W.; von Appen, J.; Choi, P.; Dronskowski, R.; Raabe, D.; Bleck W.: Atomic-scale investigation of ε and θ precipitates in bainite in 100Cr6 bearing steel by atom probe tomography and ab initio calculations, Acta Materialia 61 (2013a), 7582–7590

Song, W.; Rong, J.; Prahl, U.; Bleck, W.: Modelling of bainitic transformation kinetics under continuous cooling in forging steel 30MnCrB, 7th Int. Conf. on Physical and Numerical Simulation of Materials Processing, 16.-19.6.2013b, Oulu, Finland

Steinbach, I.: Phase-field models in materials science, Modelling and Simulation in Materials Science and Engineering; IOP PUBLISHING LTD, 17, 073001–31, (2009)

Takahashi, M.; Bhadeshia, H. K. D. H.: Mater. Sci. Tech., 1995, Vol. 11, pp. 874–881

Theiry, D.; Bettinger, R. et al., Aluminiumfreier Wälzlagerstahl, Stahl und Eisen 117 (1997) 8, 79–89

v. Appen, J. et al.: SFB 761—Stahl—ab initio; Quantenmechanisch geführtes Design neuer Eisenbasiswerkstoffe; Jahresmagazin Ingenieurwissenschaften (2009), 132–134

Wang, J.; Van der Wolk, P.J,; Van der Zwaag, S.: Journal of Materials Science 35, 2000, pp. 4393–4404

Zhang, L.; Thomas, B. G.: State of the art in evaluation and control of steel cleanliness, ISIJ International 43 (2003) 3, 271–291

Part IV
Integrated Technologies

B. Lauwers, G. Hirt, C. Hopmann, C. Windeck

To enable an economically reasonable production in high wage countries it is always a major challenge to increase the productivity of the processes. This can be achieved by getting properties out of the used materials or by reducing the material use while keeping the properties. To be successful in these paths a high knowledge of process and material is necessary, which can be a differentiating factor for producing companies. This implies that different disciplines need to be brought together in the research and development phase.

One way to increase productivity is to generate new integrated/hybrid production processes by combining several steps, respectively, by adding assistance within the main production step. For the latter, even synergetic effects may be exploited. The first contribution in this part describes integrated metal manufacturing processes for both adding and combing processes. It shows with many examples how integrated technologies help to increase accuracy and productivity.

An economically reasonable production is also given when customers' demands for individual products in small lot sizes can be satisfied fast and further functionalities can be integrated easily. Representatively, the second contribution of this part shows the incremental sheet forming (ISF) and its evolution from the basic process to new process variants with added as well as combined steps: By adding a laser system the metal is locally heated up prior to the forming so that the processing speed is increased. By e.g. integrating stretch forming and ISF different functionalities can be realized in one clamping. Again, several examples show the general benefit of integrated processes.

Besides the integration of processes also the integration of different material types leads to results that one material alone could not fulfil. Plastics show among others good insulation, specific mechanical properties and lightweight characteristics. Metals feature high overall mechanical properties over a wide temperature range and high thermal and electrical conductivity. In a combination of these materials hybrid products can be made. The development of automated one-step processes is a major topic in the third contribution. Focus is laid on the combination of injection moulding and die casting with special emphasis of electrically conductive parts.

Chapter 8
Productivity Improvement Through the Application of Hybrid Processes

Bert Lauwers, Fritz Klocke, Andreas Klink, Erman Tekkaya, Reimund Neugebauer and Donald McIntosh

Abstract Many parts require high strength materials, exhibiting high temperatures or where formability should be reduced, requiring new processing technologies. The application of hybrid manufacturing processes can answer the needs. This paper gives first a classification of hybrid manufacturing processes, followed by a description of various productivity improvements. The latter is also demonstrated by various examples in cutting, grinding, forming and chemical & physical processes like EDM, ECM and laser.

8.1 Introduction

Advanced mechanical products such as gas turbines, aerospace & automotive parts and heavy off-road equipment, rely more and more on advanced materials to achieve required performance characteristics. Many parts require high strength materials, exhibiting high temperatures or where formability should be reduced, requiring new processing technologies. The application of hybrid manufacturing processes can answer the needs.

B. Lauwers (✉)
Department of Mechanical Engineering, KU Leuven, Celestijnenlaan 300, Box 2420, 3001 Louvain, Belgium
e-mail: bert.lauwers@kuleuven.be

F. Klocke · A. Klink
Laboratory for Machine Tools and Production Engineering, RWTH Aachen University, Aachen, Germany

E. Tekkaya
Institute of Forming Technology and Lightweight Construction (IUL) at TU Dortmund University, Dortmund, Germany

R. Neugebauer
Fraunhofer Institute for Machine Tools and Forming Technology IWU, Chemnitz, Germany

D. McIntosh
Pratt & Whitney Canada Corporation, Brossard, QC, Canada

© The Author(s) 2015
C. Brecher (ed.), *Advances in Production Technology*,
Lecture Notes in Production Engineering, DOI 10.1007/978-3-319-12304-2_8

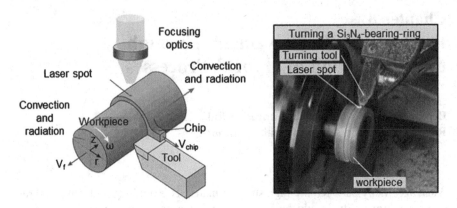

Fig. 8.1 Principle and application of laser assisted turning (Shin 2011)

According to the international academy for production engineering (CIRP), hybrid manufacturing processes are based on the simultaneous and controlled interaction of process mechanisms and/or energy sources/tools having a significant effect on the process performance (Lauwers et al. 2014). The wording "simultaneous and controlled interaction" means that the processes/energy sources should interact more or less in the same processing zone and at the same time.

Two distinct examples of hybrid processes are given to better explain what a hybrid process means. First, laser assisted cutting, where the laser beam is directly focused in front of the cutting tool, resulting in easier machining and higher process performance (Fig. 8.1).

In this process, the main material removal mechanism is still the one occurring in conventional cutting, but the laser action softens the workpiece material, so machining of high alloyed steels or some ceramics becomes easier. It is only by applying the laser energy and the mechanical cutting energy at the same time that more efficient machining becomes possible. Due to the softening effect, the process forces decrease drastically and often better surface quality can be obtained.

A second example in the area of forming is curved profile extrusion (CPE) (Klaus et al. 2006), where extrusion and bending is combined within a unique new process. In comparison to the traditional processing route for manufacturing of curved profiles (Fig. 8.2), where first the straight profile is extruded and then in a second process bended, in CPE, the extruded profile passes through a guiding tool, moveable by a linear axes system, naturally bending the profile during extrusion.

Thus, the material flow in the extrusion die is influenced by the superimposed bending moment of the guiding tool and the additional friction force in the bearing areas. Consequently, the material is accelerated at the outside and decelerated at the inside of the profile so that a controlled curvature results from this differing material flow. Due to the bending during extrusion within the die, this new forming process causes no cross-sectional distortion of the profile, no spring back, and nearly no decrease in formability. Compared to warm bending tests, process forces could be

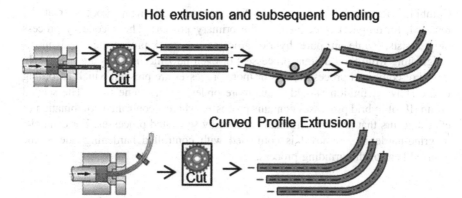

Fig. 8.2 Comparison of traditional manufacturing of curved profiles versus curved profile extrusion (hybrid)

drastically reduced to 10–15 % of the bending force that would be required if only warm bending would have been applied.

8.2 Classification of Hybrid Processes

Figure 8.3 gives a further classification or grouping of hybrid processes and some examples. The first group (I) contains processes where two or more energy sources/tools are combined and have a synergetic effect in the processing zone. A further classification is made in "Assisted Hybrid Processes" (I.A) and "Mixed or

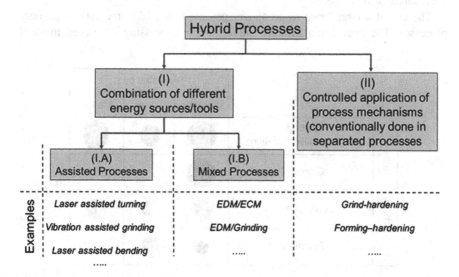

Fig. 8.3 Classification of Hybrid Processes

Combined Processes" (I.B). In assisting processes, a main process (material removal, forming,....) is defined by the primary process. The secondary process only assists, while in pure hybrid processes, several processing mechanisms (originating from the different processes) or even new mechanisms are present. In mixed or combined processes, two or more processes are present, which according to the above definition should occur more or less at the same time. The second group (II) of hybrid processes contains processes where a controlled combination of effects occurs that are conventionally caused by separated processes. For example, in grind-hardening, removal is combined with controlled hardening due to the induced heat of the grinding process.

8.3 Assisted Hybrid Processes

Figure 8.4 shows common combinations of a primary process with a secondary assisting process to create a hybrid assisted process technology. It can be concluded that the most important secondary assisting processes are vibration (ultrasonic), laser and media (fluid) assistance. Vibration assisted technologies are used in various primary processes to support the material removal. In these processes, a small vibration (average amplitudes: 1...200 μm, frequencies: 0.1...80 kHz) is added to the tool or workpiece movement. In most systems, especially in cutting and grinding operations, the amplitudes are in the range of 1–15 μm and vibration is within a frequency range from 18 to 25 kHz and the vibration itself is generated by piezoelectric elements within the tool holder, spindle or workpiece holding system. Therefore, the term "Ultrasonic Assisted Machining" (US) is also often used for these kinds of processes.

The use of a laser beam as secondary process is available for various primary processes. The laser beam strongly influences the processing zone (e.g. material

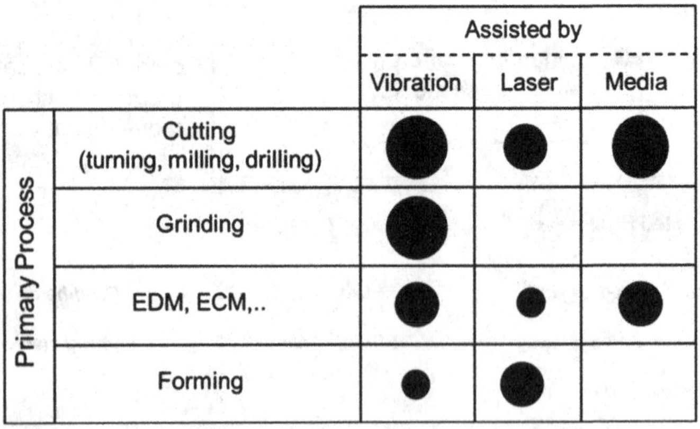

			Assisted by		
			Vibration	Laser	Media
Primary Process		Cutting (turning, milling, drilling)	●	●	●
		Grinding	●		
		EDM, ECM,..	●	●	●
		Forming	●	●	

Fig. 8.4 Combinations of assisted hybrid processes

softening in cutting, changing electrolyte conditions in ECM, material elongation and bending in forming, etc.) so processing/shaping/machining becomes easier.

The third very important group of secondary assisting processes incorporates the so called "Media-assisted Processes". This includes high pressure and cryogenic cooling/lubrication applied by dedicated jets or cooling nozzle systems. It is also used in forming (e.g. the pneumo-mechanical deep drawing process), where a pressurized medium is used to pre-stretch the sheet during the conventional deep drawing process. The borderline to conventional cooling and lubrication applications is not always clearly defined but it can be stated that there must be a significant process improvement initiated by the media assistance. Assisted hybrid processes results in a number of strong positive effects on the process performance. Often, the term "1 + 1 = 3" is used, meaning that the positive effect of the hybrid process is more than the double of the advantages of the single processes. In the following sections, a number of productivity improvements are described and demonstrated by one or two examples (Schuh et al. 2009).

8.3.1 Reduction of Process Force

Various assisted hybrid processes, such as laser assisted turning and ultrasonic assisted grinding show strong process force reductions. An example for major reduction of the drilling torque to a quasi-static value is shown in Fig. 8.5 (left) for the vibration assisted deep hole drilling of electrolytic copper ECu57 (Heisel et al. 2008). A virtually constant value can be reached independent of feed rate for no-load vibration amplitudes (A) in the range of about 5–10 μm. Also a better chip breakage for the ductile material was achieved.

Figure 8.5 (right) shows the results of a superimposing oscillation in sheet bulk metal forming (Behrens et al. 2013). The part itself is manufactured by deep drawing and the gearing is produced by bulk forming in a combination with superimposing oscillation, which leads to significantly reduced process forces. The investigations showed that with increasing process requirements such as lower die clearance, the superimposed oscillation has a greater effect on the reduction of the forming force and the spring back behaviour.

8.3.2 Higher Material Removal Rate

Not only an increase of the feed rate (due to lower process forces), but also the interaction effect between the two energy sources often results in higher material removal rates. Figure 8.6 shows the High-Speed Electro-erosion Milling (HSEM) process where high material removal rates are achieved by promoting controlled electric arcing through the use of a non-dielectric medium and a spinning electrode, enabling simultaneous multiple discharges and arcs to occur. The process starts

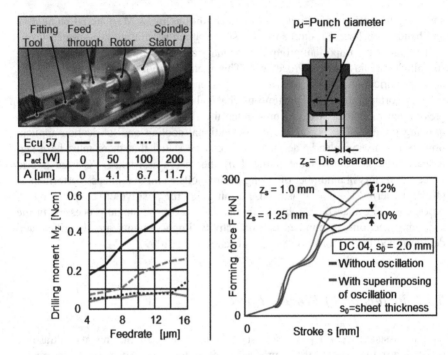

Fig. 8.5 Process force reductions in vibration assisted hole drilling (*left*) and sheet bulk metal forming (*right*)

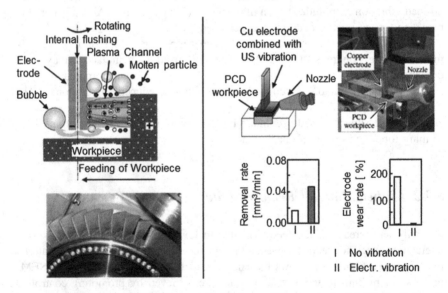

Fig. 8.6 Increase of material removal rate for High-speed Electro-erosion Machining (*left*) and vibration assisted EDM (*right*)

with multiple ionic micro-bridges in the small gap. The applied voltage triggers gas bubble generation and breakdown as well as instantaneous short-circuiting, resulting in rapid metal erosion in many locations. A metal removal rate of approx. 200 cm³/min has been achieved with a 25 mm thick disk electrode. During blisk milling of Inconel 718 the process achieves a 3 times higher material removal rate compared to conventional cutting (Wei et al. 2010).

In vibration assisted EDM, an additional relative movement is applied in the system tool electrode, workpiece and dielectric fluid in order to increase the flushing efficiency, resulting in a higher material removal rate and better process stability (Fig. 8.6, right) (Ichikawa and Natsu 2013). In addition, the tool wear is drastically reduced by applying tool/workpiece vibration.

8.3.3 Reduced Tool Wear

Besides vibration assisted EDM processes, tool wear reductions are also observed in other assisted hybrid processes. Figure 8.7 (left) shows the advantageous effects on tool life in cryogenic machining of TiAl6V4 (media assisted process). The cryogenic cooling with liquid nitrogen LN2 or carbon dioxide CO_2 is widely applied for machining of Ni- and Ti-based superalloys.

Also in the case of vibration assisted turning, the periodic disengagement of the cutting tool during vibration assistance offers the opportunity for ultra-precision machining of hardened steel, glass and even other ceramic materials with single crystal diamond tools with reduced process forces and increased surface qualities

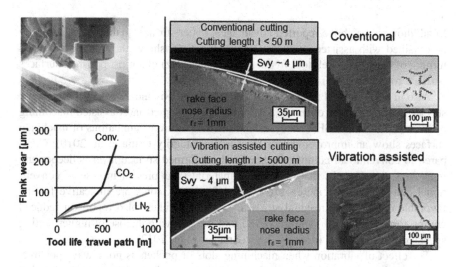

Fig. 8.7 Cryogenic cooling (milling set-up) of TiAl6V4 (*left*) and tool wear behaviour in vibration assisted turning (*right*)

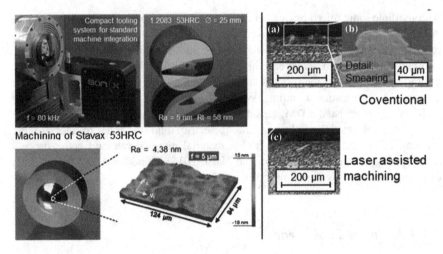

Fig. 8.8 Excellent surface quality in Vibration assisted turning (*left*) and laser assisted machining (*right*)

(at least for ferrous materials) (Bulla et al. 2012). The machining results in drastically reduced tool wear achieving highest geometrical accuracies (Fig. 8.7, right). Also the chip breakage can be positively influenced by the vibration assistance resulting in favourable short or even discontinuous chips.

8.3.4 Excellent Surface Quality

In addition to the above reported advantages, better surface qualities can generally be obtained with assisted processes. Figure 8.8 (left) shows the vibration assisted turning of hardened steel (tool frequency op to 80 kHz) of almost polished surfaces (surface roughness (Ra) values within the nm-range).

Also in laser assisted machining of difficult-to-cut Ni- and Ti-based alloys, better surface qualities can be obtained. Figure 8.8 (right) depicts laser assisted machining of Inconel 718 where SEM analyses and microstructure examinations of machined surfaces show an improvement of the surface integrity (Attia et al. 2010). Compared to conventional cutting, the plastically deformed surface layer is deeper and more uniform. The absence of smeared material (was present in case of conventional cutting) and the increased plastic deformation zone, are indicative for the favorable compressive residual stresses. Other researchers also report on reduced cutting forces (40–60 %) and improved tool life allowing the use of higher cutting speeds during LAM of Inconel alloys (Brecher et al. 2012).

The effect of vibration when machining slots or pockets is not always positive. For example in rotary ultrasonic assisted grinding (Fig. 8.9), a tool vibration perpendicular to the feed direction can result in surface cracks due to the

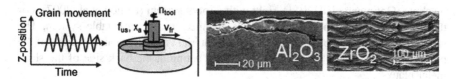

Fig. 8.9 Negative effect of the vibration on the surface quality in rotary ultrasonic assisted

hammering of the tool (see for example the machining of a Al_2O_3 (Vanparys 2012)). In general, surface textures of machined hard materials by vibration assisted grinding show mixed material removal mechanisms (MRM): plastic deformation and brittle removal. The type of material removal mechanism depends on the material properties, the amplitude of vibration and the machining parameters. Brittle removal certainly increases the material removal rate, additionally supporting lower process forces, but should be avoided in finishing processes.

Also in the case of machining of stabilized ZrO_2, the effect of the vibration is clearly visible on the surface texture (Lauwers et al. 2010). As ZrO_2 is a tough material (among all ceramic materials), no brittle removal is observed, but the impact of the grain movement is clearly visible.

8.3.5 High Precision

As process force reductions is a main advantage, hybrid assisted processes find interesting applications in micro machining such as micro EDM and micro laser assisted milling.

But high precision is also obtained for example in laser assisted single point incremental forming (Duflou et al. 2007; Göttmann et al. 2011) (Fig. 8.10). The laser softening effect not only extends the formability limits, the spring back is also reduced. The figure shows two set-ups, left performed on a classical milling

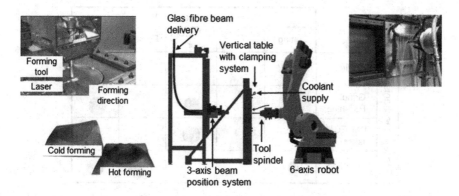

Fig. 8.10 Laser assisted single point incremental forming of sheet metal parts

machine (Göttmann et al. 2011), right with a robot system (Duflou et al. 2007). For the robot set-up, the sheet metal plate is clamped in a vertical table system. The shaping tool is moved by a robot, while the material is locally heated in front of the moving tool by a laser beam (Nd-YAG laser) acting at the back side of the sheet metal plate.

8.4 Mixed Processes and Process Mechanisms

In mixed or combined processes, two or more processes are present, which according to the definition should occur more or less at the same time. Research and development is focused on the investigation of new combinations, enhancing process performance. Figure 8.11 shows an overview of the most important process combinations.

The size of the bullets is related to the number of process combinations found in literature (Lauwers et al. 2014). Grinding and polishing is combined with EDM as well as ECM. Also the combination between grinding and hardening is coming up as an interesting hybrid process. Many process combinations exist between physical and electro-chemical processes (EDM, ECM, laser). Also in forming, processes like extrusion, spinning, bending are often combined to increase the process performance. In general, processes are combined to enhance advantages and to minimize potential disadvantages found in an individual technique (Rajurkar et al. 1999).

8.4.1 Combinations with EDM

In the area of process combinations with EDM the integration of grinding and spark erosion processes has gained an important role (Kozak et al. 2002). Figure 8.12 (left) shows the basic principle and the application of a EDM-Grinding hybrid

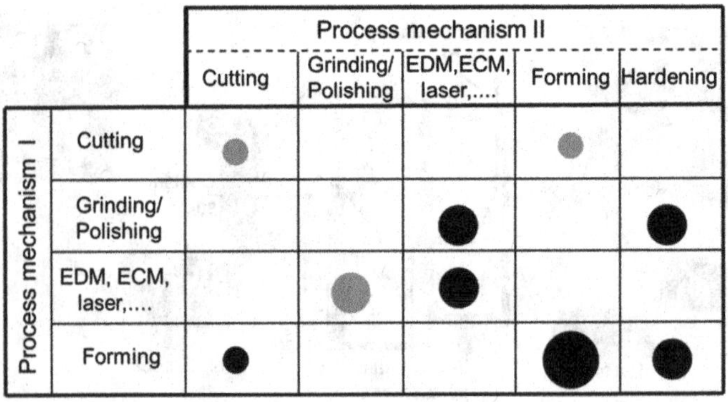

Fig. 8.11 Mixed processes and process mechanisms—possible combinations

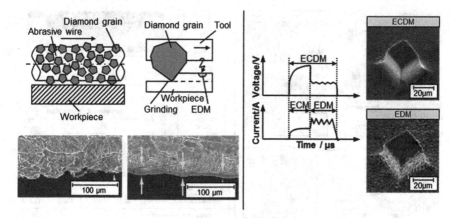

Fig. 8.12 Combinations with EDM: with grinding (*left*), with ECM (*right*)

processes, more specifically Abrasive-Wire-EDM where enhanced material removal is realized by the synergy between spark erosion and abrasion.

During EDM grinding of difficult-to-machine but electrically conductive materials like cemented carbides with metal bonded diamond grinding wheels, the grinding performance is enhanced by both effectively removing material from the workpiece and declogging the grinding wheel surface.

The combination of ECM and EDM has been widely investigated by many researchers (Lauwers et al. 2014). According to the principle of ECDM (Fig. 8.12, right), the discharge delay time of the EDM process is used for electrochemical based broad surface abrasion followed by local thermal material removal in consequence of discharge formation. By adjusting the process parameters smooth surface finishes with reduced thermally influenced rim zones and high geometrical precision can be achieved during machining of micro features.

8.4.2 Combinations with Grinding

The hybrid process combination of grinding and ECM (Fig. 8.13, left) was already developed in the 1960s in order to get a high-efficient and burr-free material removal process for difficult-to-machine aerospace alloys and cemented carbides (Becker-Barbrock 1966). This hybrid process allows for example the burr-free grinding of honeycomb structures for turbine applications (Fig. 8.13, left). Process variants were also developed for ECM-honing applications (Scholz 1968). Nowadays, alternative technologies have been developed, largely reducing the application of this process because of the high complexity of process control and environmental concerns. Figure 8.13 (right) shows another application, the precise machining of small holes (Zhu et al. 2011).

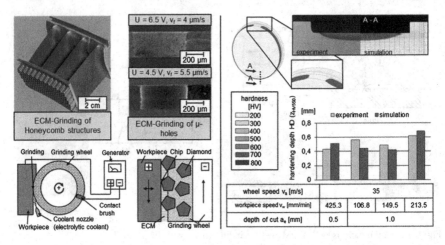

Fig. 8.13 Combinations with grinding: grinding and ECM (*left*), grinding and hardening (*right*)

The grind-hardening process utilizes the induced heat of the grinding process for local surface hardening on the workpiece. For achieving the high heat input rate the grinding process is applied with higher depth of cut and slow feed speeds. For the process combination the additional hardening process and the logistics are completely eliminated saving time, energy and production costs (Zäh et al. 2009). Figure 8.13 (right) shows the results of hardness measurements as reported in Kolkwitza et al. (2011).

8.4.3 Process Combinations with Hardening

Besides grind-hardening, there are other processes which are combined with hardening. An example is the hot stamping process used for the manufacturing of high strength components for lightweight construction (Karbasian and Tekkaya 2010). Within the direct hot stamping process (Fig. 8.14, left) an aluminum–silicon coated blank is heated up above the Ac3-temperature of the material and dwelled for a certain time to ensure a homogeneous austenitic microstructure. Afterwards, the blank is transferred to a press in which it is formed and simultaneously quenched by tool contact. With cooling rates above 27 K/s the commonly used boron-manganese steel 22MnB5 develops a martensitic microstructure with an ultimate tensile strength of 1500 MPa and an ultimate elongation of 5–6 % (Lechler 2009). The hybrid character is given since the quenching of the workpiece material is applied in the calibration phase of the hot forming operation which leads to reduced springback. The combination of forming and hardening makes 22MnB5 steel an ideal solution for the construction of structural elements and safety-relevant components in the automotive industry, in particular in view of the implementation of

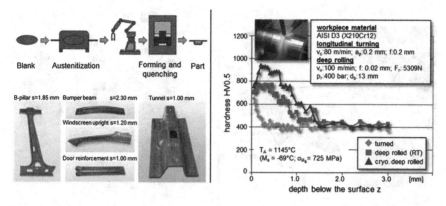

Fig. 8.14 Other process combinations with hardening: hot stamping and hardening (*left*), deep rolling and hardening (*right*)

penetration protection in the areas of the passenger cabin or motor (N.N. 2008). Figure 8.14 (left) also shows some automotive applications of hot stamping: A-pillars, B-pillars, side impact protections, frame components, bumpers, bumper mounts, door pillar reinforcements, roof frames, tunnels, rear and front end cross members. The sheet thickness in these parts varies between 1.0 and 2.5 mm.

Another hybrid processes combining with hardening is surface hardening by cryogenic deep rolling (Meyer et al. 2011). In this hybrid process (which could be seen as a kind of media assisted process), workpieces are exposed to the mechanical loads of a deep rolling process and a cryogenic treatment cooling applying CO_2-snow simultaneously. The hybrid process causes plastic deformation and strain induced martensitic transformations into depths of up to 1.5 mm (Fig. 8.14, right).

8.4.4 Combination of Forming Processes

Some examples of combinations of forming processes are presented in Fig. 8.15. The first process is a combination of a tube spinning and a tube bending process (Fig. 8.15, left) (Becker et al. 2012). A tube is being clamped on a feeding device and is transported through a sleeve to the spinning tool. The three spinning rolls of the spinning tool are rotating around the tube at a defined rotational speed. The spinning process creates a diameter reduction of the tube. To manufacture a bent structure a freeform bending process is superposed. Due to this process setup the production of bent structures can be realized with variable tube diameters. In this hybrid process, the spinning process significantly influences the bending results, which is shown by reduced process forces and reduced springback. Figure 8.15 (left) also shows a prototype machine and industrial manufactured samples. Tube diameters up to 90 mm can be processed as well as tube lengths of 3000 mm. Also the bending of three dimensional parts is possible due to a change of the bending plane by rotation of the pusher device.

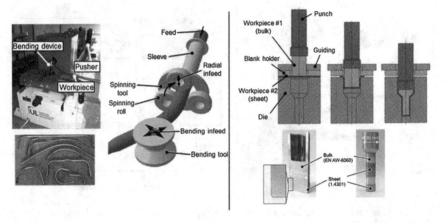

Fig. 8.15 Other combinations of forming processes: extrusion and spinning (*left*), deep drawing and cold forging (*right*)

The combined process of deep drawing and cold forging is a new hybrid metal forming process to produce composite products from different combinations of materials (Jäger et al. 2012). As presented in Fig. 8.15 (right), a one side coated circular sheet is positioned centrally above the contour-shaping die. The opening of the die has a small radius, which serves as a drawing edge (die radius). By substituting the deep drawing mandrel by a cylindrical bulk metal workpiece, the sheet is deep drawn into the shape of a cup which partly covers the bulk component. With increasing stroke the bulk metal workpiece starts to be cold forged, while the sheet component is additionally formed or even calibrated. At the end of the cold forging process, the punch moves upwards and the workpiece is pressed out by an ejector from the bottom of the tool. Depending on the diameter of the sheet in relation to the height of the bulk part, there is a partial or a complete cladding of the component. Composite metal structures with a cold forged bulk material in the core partly covered with a deep-drawn sheet material can be produced (Fig. 8.15, right). It is expected that the use of a bulk part instead of a conventional mandrel allows a greater drawing ratio because of the simultaneous movement and deformation of the sheet and the bulk part. Furthermore, due to the cold forging process an additional reduction of the cross section can be carried out.

8.5 Conclusions

This paper gave a brief overview of advanced manufacturing through the implementation of hybrid processes. The process combinations are used to considerably enhance advantages and to minimize potential disadvantages found in individual techniques. Within hybrid production processes different forms of energy or forms of energy caused in different ways are used at the same time at the same zone of impact.

The combination of processes result various advantages that often occur at the same time: lower processes forces, higher precision, higher productivity,... The development of hybrid processes is continuously evolving, from basic development towards industrial implementation. Further developments are driven on the one hand by industrial needs to manufacture highly engineered mechanical products made of advanced materials and on the other hand to process parts in a more productive and energy efficient way.

Acknowledgments The authors would like to thank all CIRP members who have contributed to the CIRP collaborative working group on "Hybrid Processes". In addition, the authors would like to thank R. Wertheim (TU Chemnitz), M. Kuhl (TU Chemnitz), A. Rennau (TU Chemnitz), A. Jäger (TU Dortmund), J. Bouquet (KU Leuven) and H. Romanus (KU Leuven) for their valuable input in the preparation of this paper.

References

Attia H., Tavakoli S., et al (2010), Laser-assisted high-speed finish turning of superalloy Inconel 718 under dry conditions, CIRP Annals—Manufacturing Technology 59/1:83–88

Becker-Barbrock, U. (1966), Untersuchung des elektrochemischen Schleifens von Hartmetall und Schnellarbeitsstahl, Dissertation RWTH Aachen

Becker C., Staupendahl D., et al (2012) Incremental Tube Forming and Torque Superposed Spatial Bending—A View on Process Parameters, Proceedings of the 14th International Conference "Metal Forming 2012", steel research international, 159–162.

Behrens B., Hübner S., et al (2013), Influence of superimposed oscillation on sheet-bulk metal forming, Key Engineering Materials, 554–557:1484–1489

Brecher C., Emonts M., et al (2012), Laserunterstützte Fräsbearbeitung, wt Werkstattstechnik online, 102,6: 353–356

Bulla B., Klocke F., et al (2012) Influence of different steel alloys on the machining results in ultrasonic assisted diamond turning, Key Engineering Materials, 523–524:203–208

Duflou J.R., Callebaut B., et al (2007) Laser Assisted Incremental Forming—Formability and Accuracy Improvement, CIRP Annals—Manufacturing Technology, 56/1:273–276

Göttmann A., Diettrich J., et al (2011) Laser-assisted asymmetric incremental sheet forming of titanium sheet metal parts, Prod. Eng. Res. Devel., 5:263–271

Heisel U., Wallaschek J., et al (2008), Ultrasonic deep hole drilling in electrolytic copper ECu 57, CIRP Annals—Manufacturing Technology, 57:53–56

Ichikawa T., Natsu W., (2013) Realization of micro-EDM under ultra-small discharge energy by applying ultrasonic vibration to machining fluid, Procedia CIRP, 6:326–331

Jäger A., Hänisch S., et al (2012) Method for producing composite parts by means of a combination of deep drawing and impact extrusion, Patent application PCT/DE2011/00001053.

Karbasian H., Tekkaya A.E. (2010) A review on hot stamping, Journal of Materials Processing Technology, 210/15:2103–2118

Klaus A., Becker D., et al (2006) Three-Dimensional Curved Profile Extrusion—First Results on the Influence of Gravity, Advanced Materials Research, 10:5–12

Kolkwitza B., Foeckerer T., et al (2011), Experimental and Numerical Analysis of the Surface Integrity resulting from Outer-Diameter Grind-Hardening, Procedia Engineering 19 (2011) 222–227

Kozak J.; Rajurkar K.P.; et al (2002) Self-Dressing and Performance Characteristics in Rotary Abrasive Electrodischarge Machining, Transac. of the North Amer. Manuf. Res. Institution of SME, Vol.30, (2002), p. 145–152

Lauwers B., Klocke K., et al (2014) Hybrid processes in manufacturing. CIRP Annals—Manufacturing Technology, 63:561–583

Lauwers B., Bleicher F., et al (2010) Investigation of the process-material interaction in ultrasonic assisted grinding of ZrO_2 based ceramic materials, 4th CIRP International Conference on High Performance Cutting 2:59–64

Lechler J. (2009) Beschreibung und Modellierung des Werkstoffverhaltens von presshärtbaren Bor-Manganstählen, Meisenbach (Verlag), 978-3-87525-286-6 (ISBN)

Meyer D., Brinksmeier E., et al (2011) Surface hardening by cryogenic deep rolling, Procedia Engineering, 19:258–263

N. N. (2008), Steels for hot stamping—very high strength steels, ArcelorMittal

Rajurkar K.P., Zhu D., et al (1999) New Developments in Electro-Chemical Machining, CIRP Annals—Manufacturing Technology 48/2:567–579

Scholz E. (1968) Untersuchung des elektrochemischen Honens, Dissertation RWTH Aachen.

Schuh G., Kreysa J., et al (2009), Roadmap "Hybride Produktion", Zeitschrift für Wirtschaftlichen Fabrikbetrieb, 104/5:385–391

Shin Y. (2011) Laser assisted machining, www.industrial-lasers.com/articles, 26/1.

Vanparys M. (2012) Ultrasonic Assisted Grinding of Ceramic Components, PhD thesis, KU Leuven, ISBN 978-94-6018-600-4

Wei B., Trimmer A.L., et al (2010) Advancement in High Speed Electro-Erosion Processes for Machining Tough Metals, Proceedings of the 16th International Symposium on Electromachining, 193–196

Zäh M.F., Brinksmeier E., et al (2009) Experimental and numerical identification of process parameters of grind hardening and resulting part distortions, Prod. Eng. Res. Devel. 3:271–279

Zhu D., Zeng Y.B., et al (2011) Precision machining of small holes by the hybrid process of electrochemical removal and grinding, CIRP Annals—Manufacturing Technology, 60/1:247–250

Chapter 9
The Development of Incremental Sheet Forming from Flexible Forming to Fully Integrated Production of Sheet Metal Parts

Gerhard Hirt, Markus Bambach, Wolfgang Bleck, Ulrich Prahl and Jochen Stollenwerk

Abstract Incremental Sheet Forming (ISF) was devised as a flexible forming process in the 1990s. The basic principle of ISF is that a generic forming tool moves along a tool path and progressively forms a metal sheet into the desired shape. The tool is either moved using CNC machines or industrial robots. Applying CNC technology or robots to sheet metal forming allows for replacing expensive dedicated tooling and for a fast transfer from the CAD model to the formed part. Since its first applications in the 1990s ISF has undergone tremendous developments. Various process variants such as double-sided ISF and hybrid process combinations such as heat-assisted ISF as well as stretch-forming and ISF have been put forward. The present contribution gives an overview of these developments with a special focus on the outcome of the research accomplished within the cluster of excellence "Integrative Production Technology for High Wage Countries", where the development of fully integrated sheet metal production facilities is envisioned as the next evolution step of ISF. The development of dedicated equipment for hybrid and fully integrated sheet metal manufacturing and specialized CAX environments as well as applications are described to show the potential of the technology.

G. Hirt (✉) · M. Bambach
Metal Forming Institute (IBF), RWTH Aachen University,
Intzestr. 10, 52056 Aachen, Germany
e-mail: hirt@ibf.rwth-aachen.de

W. Bleck · U. Prahl
Department of Ferrous Metallurgy (IEHK), RWTH Aachen University,
Intzestr. 1, 52072 Aachen, Germany
e-mail: bleck@iehk.rwth-aachen.de

J. Stollenwerk
Fraunhofer Institute for Laser Technology ILT, Steinbachstr. 15,
52074 Aachen, Germany
e-mail: jochen.stollenwerk@ilt.fraunhofer.de

© The Author(s) 2015
C. Brecher (ed.), *Advances in Production Technology*,
Lecture Notes in Production Engineering, DOI 10.1007/978-3-319-12304-2_9

9.1 Introduction to Incremental Sheet Metal Forming

Incremental sheet forming (ISF) is a flexible forming process for small batch manufacturing and rapid prototyping of almost arbitrary 3D shapes. In ISF, a clamped sheet metal is progressively formed by a moving forming tool (Fig. 9.1, right). In contrast to conventional sheet metal forming processes such as deep drawing (Fig. 9.1, left), only a single die is needed, which does not have to be a full male or female die but can be a partial support.

The tool path covers the surface of the desired product, similar to the finishing stage in z-level machining. In every instant of the forming process in which the tool moves over the sheet metal, localized plastic deformation is produced and the final part shape is the result of all localized plastic deformation events. Several variants of the incremental sheet forming process have been developed in the past:

- *Conventional ISF*. Conventional ISF comprises the variants of 'single point incremental forming' (SPIF) and 'two-point incremental forming' (TPIF). In SPIF, either no support at all or only a simple rig is used to support the outer contour of the part. In TPIF, the sheet metal is formed over a full or partial positive die.
- *Double-sided ISF*. In this process variant, a tool is used on either side of the sheet, with one tool acting as the master forming tool and the other one acting as a local support. This process was investigated by Meier et al. (2007), Maidagan et al. (2007) as well as Malhotra et al. (2011).
- *Stretch-forming and ISF*. To overcome some of the limitations of conventional ISF process variants such as the long process time, the pronounced thinning and

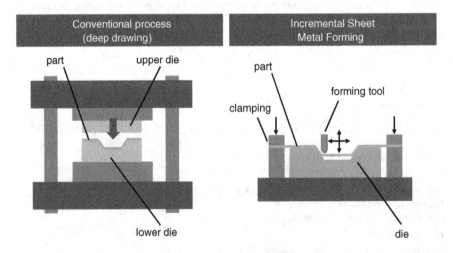

Fig. 9.1 Illustration of conventional deep drawing and 'single point incremental forming' (SPIF)

the limited geometrical accuracy, ISF was combined with stretch-forming to a hybrid process by Taleb Araghi et al. (2009). The process was performed on a dedicated machine that combines four stretch-forming units and a CNC unit for ISF.

- *Heat-assisted ISF*. In order to form e.g. titanium or magnesium alloys with a low formability at room temperature, heat-assisted ISF variants were developed, such as laser-assisted ISF by Duflou et al. (2007) and Göttmann et al. (2011), or ISF with resistance heating, see e.g. Göttmann et al. (2012). These variants are also hybrid processes, whose set-up and control is much more involved than for conventional ISF.

Both conventional ISF and the newer process variants have been developed with great effort by a number of research groups, but up to now only with limited industrial take-up. The main limitations of conventional ISF are (i) the limited geometrical accuracy, (ii) excessive sheet thinning, (iii) the long process time and (iv) the need for dedicated CAE tools. Besides that, potential markets for ISF are sheet metal parts made from titanium and magnesium alloys which are hard to form at room temperature and require forming at elevated temperature.

To meet the above-mentioned challenges and to make ISF viable in an industrial context, various technological developments beyond conventional ISF are necessary. This contribution gives an overview of recent developments in ISF with a strong focus on hybrid ISF processes developed in the cluster of excellence "integrative production technology for high-wage countries".

The paper is organized as follows: The next section gives an overview of the design of the dedicated machine for hybrid ISF processes and the CAX tools needed to operate the machine. The capabilities of the hybrid processes of stretch forming and ISF as well as laser-assisted ISF are demonstrated using case studies. Finally, the benefit of the hybrid processes compared to standard ISF is summarized.

9.2 Design of a Machine for Hybrid ISF

9.2.1 Basic Set-up for Stretch-Forming and ISF

The machine shown in Fig. 9.2 is based on a standard milling machining center into which four dedicated stretch-forming modules were integrated (shown in blue in Fig. 9.2). Thus, the process steps of milling of the die, stretch forming, ISF and trimming can be carried out on a single machine.

Particularly noteworthy is the stiff machine bed (Fig. 9.2, right), which must bear the extremely high process forces exerted by stretch-forming (about 200 kN per element) in a confined space. Furthermore, an interface has been created in the 5-axis head which can receive forming tools for ISF as well as conventional milling tools. For ISF a force limiter was developed to protect the linear and rotary axes of the milling machine from overloading. Both for stretch forming as well as for ISF a

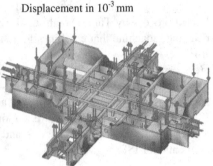

Displacement in 10^{-3} mm

Fig. 9.2 (*Left*) Hybrid machine enabling stretch forming and ISF. (*Right*) simulation of the deflection of the machine bed

workpiece holder is required, which must transmit the high process forces. The stretch-forming modules allow movements in horizontal and vertical direction and have a hinged clamp. This is necessary to allow for tangential stretch forming. The clamping of the workpieces is self-adjusting and is designed for sheet metal of 1–4 mm thickness. All movements are performed by NC controlled linear axes.

During milling of the die needed for stretch forming, the machine can be used as 5-axis milling machine with three linear axes and two rotary axes. With the system, the commonly used mold materials for ISF (aluminum, plastic, wood) can be machined.

The machine is equipped with a CNC control Siemens 840 D NCU 573 SL. Due to the flexible architecture of the controller it is possible to integrate special control functions for the stretch forming and incremental sheet forming directly in the controller. This functionality has been used to integrate a laser system as further axis (see below). The technical data of the installed system are summarized in Table 9.1.

Table 9.1 Technical data of the hybrid machine for stretch forming and ISF

Control	Siemens 840 D NCU 573 SL				
Accuracy	Positioning accuracy			Repeat accuracy	
	±0.03 mm			±0.015 mm	
Machine axes	X	Y	Z	SZ horizontal	SZ vertical
Traversing range [mm]	2.800	2.300	1.000		
Teed rate [m/min]	40	40	20		
Forming force [kN]	4	4	4	200	100
Spindle	Power	Rotation speed		Torque	Tool holder
	24 kW	18.000 U/min		38 Nm	HSK 63 A

9.2.2 Basic Set-up for Laser-Assisted ISF

To allow for localized heating, a laser optic was designed and integrated into the machine. The selected laser was a "LDF 10000" diode laser from the company Laserline. The maximum available output of 10 kW (radiation power) is sufficient to heat common sheet forming materials up to temperatures above 1000 °C. The main advantages of a diode laser are that the beam can be guided via an optical fiber. Thus, the energy required for heating can be directed right to the forming area. The movements of the forming tool can be compensated for by the optical fiber and a feed device.

Since the optical system cannot be rotated around the tool, it was designed so that the laser beam is rotated to the desired position around the tool axis. Rotation of mirrors in the laser optics causes the laser beam to move on a circle. The shape and position of the laser spot can be influenced by selecting different lenses and varying the distances between the mirror components. In the simplest version, a circular laser spot with a diameter of 35 mm is projected onto the surface of the part at a distance of 45 mm from the tool axis. The optical system described is fixed to the forming head of the hybrid machine (Fig. 9.3). The beam source used is outside the machine, so that the laser beam has to be guided to the processing point via a fiber optic. The optical system moves together with the processing head during the forming process. The laser spot is positioned by a motor that is built into the optical system.

9.2.3 CAX Environment

The CAX environment must provide suitable software tools to plan each step of the combined stretch-forming and ISF process chain as well as for laser-assisted ISF.

Fig. 9.3 Hybrid forming machine with built-in optical system

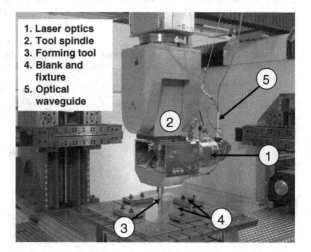

1. Laser optics
2. Tool spindle
3. Forming tool
4. Blank and fixture
5. Optical waveguide

Due to the complexity of the tool kinematics of ISF purely manual machine operation would not be possible. The same holds for stretch forming with up to 8 axes. Due to the novelty of the combination of stretch-forming and ISF process, the development of new CAM features was necessary which do not exist as standard features in common CAX systems.

The programming of the stretch-forming operation is supported by the CAM system, but it is also possible to operate the stretch-forming modules manually and to read back the trajectories into the CAM system. Previously used ISF strategies can be implemented, customized and extended. The simulation and collision checking of the forming tool, the stretch motion of the machine and the fixture situation were another important requirement.

The development of a completely new and independent CAX solution would have cost a tremendous effort. For this reason, the development of the CAX solution was carried out based on the standardized CAX platform NX from Siemens. A key criterion for the selection of NX as CAD/CAM platform is the ability to integrate own functions in the system via programming interfaces and thus to implement specific functions for stretch forming and ISF. NX offers several programming interfaces (APIs) such as NXopen (C, C++, Visual Basic).

The CAM module in NX provides basic functions for milling which can be adapted for ISF. The most important function is the "Z-level" processing. This machining strategy can be programmed with NX both in a 3-axis and with simultaneous 5-axis motion. All process steps for the production of demonstration components—geometry processing, stretch-forming, ISF and trimming the component—can be performed consistently with the developed CAX-chain.

Building on the experience gained during initial manual programming of the stretch forming units, it was possible to automate some repetitive steps. Pre-stretching of the sheet, approaching the die and bending can be combined within a single smooth trajectory. Since the stretch-forming modules move in planes, the motion can be prescribed by curves in 2d space. These can be defined separately for each stretch forming module. For the individual forming steps, the relevant parameters in the form of input values can be defined (Fig. 9.4).

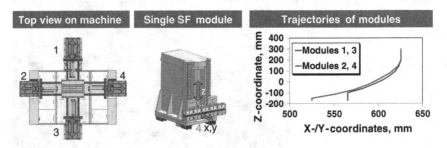

Fig. 9.4 Programming the stretch drawing (*left*) and graphical representation of the machine kinematics (*right*)

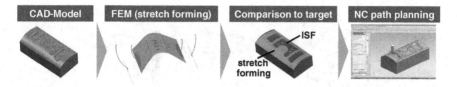

Fig. 9.5 CAD/CAM chain for hybrid stretch forming and ISF

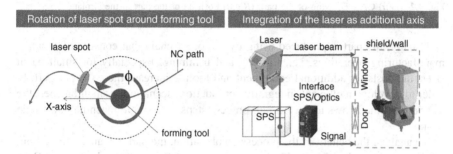

Fig. 9.6 Integration of laser as rotation axis (*left*) and integration into the machine via SPS (*right*)

Since stretch forming does not create the final geometry for most parts, the areas that still have to be formed by ISF after stretch forming have to be detected. This is accomplished by reading in results from a finite element simulation of the stretch forming process into the CAD/CAM system. The areas to be formed out by ISF are detected, and tool path planning is done only for the areas shown in red in Fig. 9.5.

The programming of the NC machine tools is often supported by a machine simulation. In the case of the hybrid process, further eight axes for the stretch-forming units are present in addition to the five axes of the forming/cutting tool. This underlines the need for system simulation in order to guarantee safe operation. The simulation avoids test runs on the system and therefore contributes significantly to shortening the process planning.

Dedicated CAM tools are also needed for laser-assisted ISF. Special laser optics were devised which guide the laser beam onto a position on the blank that is defined by a rotation angle about the X-axis (Fig. 9.6). The rotation angle is calculated in the CAM system and transferred to the forming machine along with the positioning signals for the forming tool.

9.3 Case Study: Stretch Forming and ISF

As an application part, a stiffening frame for a hydraulic access door of an AIRBUS A320 aircraft made of 1.0 mm stainless steel 1.4541 has been chosen (Fig. 9.7). The part is located on top of the pylon and allows an easy and fast maintenance operation on the hydraulic systems in this area.

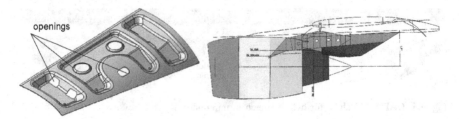

Fig. 9.7 (*Left*) CAD Drawing of the part. (*Right*) Position of the part in the airplane

Usually, the part is produced using by a process chain that consists of separate manufacturing operations, i.e., forming and trimming. Especially for small batch sizes, the costs for additional equipment and tooling increase the costs per part. In order to develop an efficient 'integrative production' scenario for small batches, the hybrid machine allows to perform all process steps on a single machine in a single set-up.

To analyze the benefit of the process combination, the part is manufactured both using conventional ISF and with the combination of ISF and stretch-forming. Both parts were trimmed to the final geometry using the milling functionality of the machine. In the case of pure ISF, the forming operation was divided into two steps. In the first step, the outer envelope of the part was formed. The pockets were manufactured separately in the second step. The forming with 'SF + ISF' took 60 min. whereas the manufacturing with the conventional ISF process needed 110 min. Using the process combination, the forming time was reduced by about 45 %.

Since the shape of the part shows smooth curvature with relatively flat pockets, low strains are expected that should not lead to significant sheet thinning but influence the geometric accuracy. Hence, within this study the dimensional accuracy was investigated. Both parts (pure ISF-part and 'SF + ISF'-part) were digitized using the gom ATOS system. The comparison of the digitized parts to the CAD model yielded the actual geometric deviations.

Figure 9.8 shows the evaluation of the geometric deviation along a longitudinal section. After trimming, pure ISF yields a lower dimensional accuracy compared to the part made by the process combination. In particular, towards the borders of the part, the geometric deviation of the section made by ISF increases strongly. It can be concluded that for the process combination 'SF + ISF', the superimposed tensile stresses due to stretch forming yield a higher dimensional accuracy in the area close to the outer borders of the part and in the transition to the flange region than pure ISF. For stretch-forming, springback compensation procedures through tool modifications could help to increase the dimensional accuracy even more.

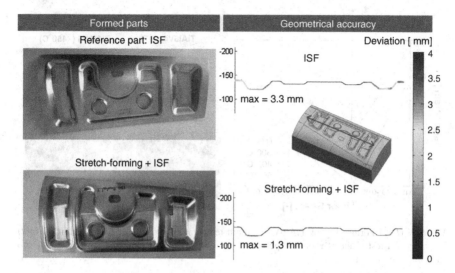

Fig. 9.8 Evaluation of geometric accuracy

9.4 Case Study: Heat-Assisted ISF

TiAl6V4 is the most commonly used titanium alloy. At room temperature, formability is very low due to the limited number of glide systems. To analyze the effect of different process parameters such as temperature, strain rate and strain on the deformation behavior of TiAl6V4, processing maps have been developed (Johnson et al. 2003). Vanderhasten et al. (2008) analyzed the deformation behavior of TiAl6V4 by uniaxial tensile testing for a wide range of strain rates and temperatures. However, since ISF is governed by complex stress states in the forming zone, processing maps or uniaxial test data are not representative. In the cluster of excellence, the formability of TiAl6V4 sheets was analyzed by recording forming limit curves, both at room temperature and for slightly elevated temperatures of 300–500 °C. This was accomplished by heating the punch to the respective forming temperature. The FLC in Fig. 9.9 on the left shows that formability increases already in the temperature range of 300–500 °C.

At such low temperatures, oxidation, i.e. the formation of the detrimental α-case, does not yet occur during laser ISF since the time spent at that temperature is too short.

Based on the analysis of formability, experiments on 1.5 mm thick sheet metals of Ti Grade 2 and TiAl6V4 were performed using local laser heating (Fig. 9.9, right). The formed geometry is a cone with a kidney-shaped base with a wall angle of 60° and a depth of 110 mm as shown in Fig. 9.8. The pitch between the z-levels was 0.35 mm. The forming velocity was set to 4000 mm/min. The settings of the laser optics were chosen such that an elliptical laser spot with dimensions 15 mm × 45 mm was projected onto the sheet metal at a distance of 45 mm to the tool. The laser power output was controlled using a closed-loop feedback controller.

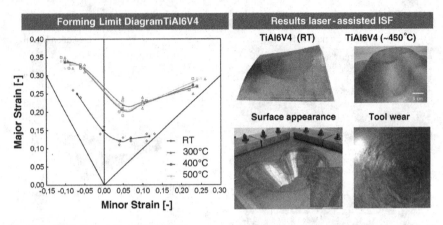

Fig. 9.9 (*Left*) Formability of TiAl6V4 at room temperature and elevated temperature. (*Right*) Forming of a test shape at room temperature and ~450 °C. Tool wear and surface quality are shown in the *bottom*

The temperature is measured in the tip of the forming tool 1 mm below the surface of the forming tool.

Although the part was formed successfully, there is excessive tool wear and the surface quality is poor. Forming of TiAl6V4 using ISF thus requires improved tool concepts such as tools with a rolling instead of a sliding contact.

9.5 Improvements by the Hybrid ISF Variants

The most restrictive process limits in conventional ISF are the geometrical accuracy and the strong dependence of sheet thinning on the wall angle of the formed part. Allwood et al. (2005) considered 28 potential sheet metal products from 15 companies to search for potential applications of ISF. A product segmentation approach revealed that only two of the 28 products comply with the capabilities of ISF. In the study, a geometrical inaccuracy of 3 mm was presumed to exist independent of part size and workpiece material. Although the assumptions made for the achievable tolerance of ISF in the product segmentation approach are oversimplified, they show that the geometrical tolerance is a key factor that decides whether a given product can be manufactured by ISF or not.

Due to the possibility to create a preform by stretch forming and due to the fact that tensile stresses can be superimposed, the combination of stretch forming and ISF helps to improve the geometrical accuracy, as illustrated in Fig. 9.10. In this illustration, it is assumed that the geometrical deviations in ISF scale with the size of the part.

Fig. 9.10 Improvement of geometrical accuracy by combining ISF with stretch forming (SF)

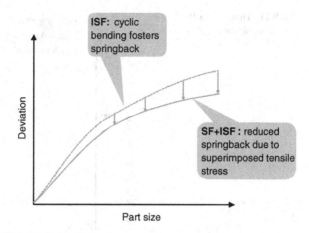

Unlike deep drawing, ISF is a process in which the sheet thickness cannot be held constant. ISF increases the surface area of the part. Thinning in the ISF process is governed by the sine law,

$$t_1 = t_0 \sin(90° - \alpha) = t_0 \cos(\alpha) \tag{9.1}$$

and increases with the wall angle. Thinning in stretch forming does not depend on the wall angle, it is rather governed by frictional constraints. The combination of ISF and SF may hence help improve the process limit determined by excessive thinning. Assuming that the sheet breaks once a certain amount of thinning is reached, forming by stretching should ideally be designed to lead to homogeneous thinning throughout the part so that there is no "weak spot" with maximum thinning. This ideal situation is hard to achieve. However, since thinning in ISF and SF affects different areas of the part, it can be complementary in many cases and hence the thickness reduction can be distributed more evenly over the part, as illustrated in Fig. 9.11. Due to volume constancy, stretching of the sheet must be compensated by thinning, i.e.

$$S_0 t_0 = S_1 t_1 \tag{9.2}$$

If the surface stretch ratio $\ln(S_0/S_1)$ is distributed unevenly over the part, there will be an area of maximum stretching and, correspondingly, maximum thinning. This area is prone to failure. To avoid it, the material should be distributed as homogeneously as possible.

The benefit of forming materials with low formability at room temperature such as titanium and magnesium alloys at elevated temperatures is shown in Fig. 9.12.

Assuming again that a maximum allowable thickness reduction exists, increasing the temperature will increase the limit strain from $\varepsilon_{max,RT}$ at room temperature

Fig. 9.11 Thinning in ISF
and the combination of SF
and ISF

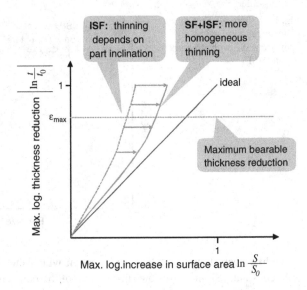

Fig. 9.12 Process limits of
ISF at room temperature and
elevated temperature

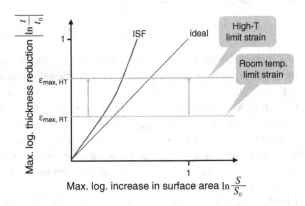

to $\varepsilon_{max,HT}$ at high-temperature deformation. This allows a larger increase in surface area and hence forming of more complex parts.

Acknowledgments The authors would like to thank the German Research Foundation (DFG) for the support of the depicted research within the Cluster of Excellence "Integrative Production Technology for High Wage Countries".

Parts of this research are funded by the German Federal Ministry of Education and Research (BMBF) within the Framework Concept "Research for Tomorrow's Production" (funding number 02PU2104), managed by the Project Management Agency Karlsruhe (PTKA).

References

Allwood J, King G, Duflou J (2005) A structured search for applications of the incremental sheet-forming process by product segmentation. Proceedings of the I MECH E Part B Journal of Engineering Manufacture 6:239–244

Duflou J, Callebaut B, Verbert J, Baerdemaeker H de (2007) Laser assisted incremental forming: formability and accuracy improvement. CIRP Annals-Manufacturing Technology 56 (1):273–276

Göttmann A, Bailly D, Bergweiler G, Bambach M, Stollenwerk J, Hirt G, Loosen P (2012) A novel approach for temperature control in ISF supported by laser and resistance heating. The International Journal of Advanced Manufacturing Technology:1–11

Göttmann A, Diettrich J, Bergweiler G, Bambach M, Hirt G, Loosen P, Poprawe R (2011) Laser-assisted asymmetric incremental sheet forming of titanium sheet metal parts. Production Engineering 5(3):263–271

Johnson AW, Bull CW, Kumar KS, Briant CL (2003) The influence of microstructure and strain rate on the compressive deformation behavior of Ti-6Al-4 V. Metallurgical and Materials Transactions A 34(2):295–306

Maidagan E, . Zettler J, Bambach M, Rodríguez P, Hirt G (2007) A new incremental sheet forming process based on a flexible supporting die system. In: Sheet Metal 2007: Proceedings of the International Conference. Trans Tech Publications Ltd, Uetikon-Zuerich, Switzerland, pp 607–614

Malhotra R, Ren F, Reddy NV, Kiridena V, Cao J, Xia ZC (2011) Improvement of geometric accuracy in incremental forming by using a squeezing toolpath strategy with two forming tools. Journal of Manufacturing Science and Engineering 133(6):61019

Meier H, Smukala V, Dewald O, Zhang J (2007) Two point incremental forming with two moving forming tools. In: SheMet '07, Proceedings of the 12th International Conference on Sheet Metal, Apr 01.-04. 2007, Palermo, Sicily, Italy. Trans Tech Publication Ltd., Switzerland, pp 599–605

Taleb Araghi B, Manco GL, Bambach M, Hirt G (2009) Investigation into a new hybrid forming process: Incremental sheet forming combined with stretch forming. CIRP Annals-Manufacturing Technology 58(1):225–228

Vanderhasten M, Rabet L, Verlinden B (2008) Ti-6Al-4 V: deformation map and modelisation of tensile behaviour. Materials & Design 29(6):1090–1098

Chapter 10
IMKS and IMMS—Two Integrated Methods for the One-Step-Production of Plastic/Metal Hybrid Parts

Christian Hopmann, Kirsten Bobzin, Mathias Weber, Mehmet Öte, Philipp Ochotta and Xifang Liao

Abstract The integration and combination of known production technologies to one-step-processes is a promising way to make existing processes more efficient and to enable more integrated products. This paper presents two integrative process technologies that are developed by the Institute of Plastics Processing (IKV) and the Surface Engineering Institute (IOT) as part of the Cluster of Excellence "Integrative Production Technologies for High-Wage Countries". In these processes, metals or metal alloys are applied to an injection moulded part, which results in a new opportunity to create electrical conductivity of plastic articles. The Integrated-Metal-Plastic-Injection-Moulding (IMKS) represents the combination of injection moulding and metal die-casting, allowing the production of plastic parts with integrated conductive tracks in one shot. The In-Mould-Metal-Spraying (IMMS) combines the injection moulding with the thermal spraying of metal. Therefore it is possible to equip electrically insulating plastic parts with metallic coatings and provide an electromagnetic shielding like cast metal parts. In the following both processes are presented and future potentials and challenges are shown.

10.1 Introduction

In traditional engineering, metals and plastics normally compete with each other (Berneck 2011; Flepp 2012). Nowadays, the growing requirements regarding functionality and complexity of parts often cannot be met by a single material. Hence the hybrid technology, which combines the advantages of different materials within

C. Hopmann (✉) · M. Weber · P. Ochotta
Institute of Plastics Processing (IKV), RWTH Aachen University,
Pontstr. 55, 52052 Aachen, Germany
e-mail: zentrale@ikv.rwth-aachen.de

K. Bobzin · M. Öte · X. Liao
Surface Engineering Institute, RWTH Aachen University,
Kackertstr. 15, 52072 Aachen, Germany
e-mail: info@iot.rwth-aachen.de

© The Author(s) 2015
C. Brecher (ed.), *Advances in Production Technology*,
Lecture Notes in Production Engineering, DOI 10.1007/978-3-319-12304-2_10

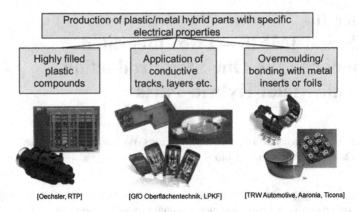

Fig. 10.1 Processes to manufacture electrically conductive parts

one part, is gaining increased importance. Amongst others the part functionality can be expanded by integrating electrical conductivity or by an improvement of the wear-resistance of the surfaces of plastics parts. Manifold applications arise in the fields of electronics as well as in the automotive industry, e.g. connectors or parts with selectively conductive areas (Drummer and Dörfler 2007).

Today, various technologies are capable of producing plastic/metal hybrid parts (Fig. 10.1).

One possible process is the injection moulding of electrically conductive polymer melts. Therefore electrical conductive materials like metal fibres, carbon black or carbon nano tubes are introduced as fillers/additives by the compounding (Pfeiffer 2005; Pflug 2005). The hybrid compounds can be processed on regular injection moulding machines; however the high filler content limits the flowability, thus requiring an adapted part design. Also the electrical properties of pure metals cannot be reached (Pfeiffer 2005).

Another way of metalising is the coating of plastic parts. For example electroplating, ion plating, chemical vapour deposition, thermal spraying or with physical vapour deposition can be used to metalise prefabricated plastic parts (Grob et al. 2003; Brosig 1996). Each process needs at least one additional process step and equipment. In addition the coating has a given thickness which leads to dimensional inaccuracy of the part. Especially the electroplating, often applied since the 1960s, is limited to special polymers, for example ABS, PC, PA (Kanani 2009; Chanda and Roy 2007).

The in-mould-assembly process comprises a preceded production of the metal parts. The prefabricated metal parts are subsequently placed in the injection mould to be overmoulded by the plastic melt. The overmoulding of metallic films is often used in terms of improving the optics and haptic (cool-touch-effect). Overmoulded grids are applied to improve the electromagnetic compatibility (EMC) of the parts. The maximum degree of deformation of the overmoulded film or grid depends on the elasticity of the used metal, which limits the geometrical freedom. In each case especially the production of the metal component is characterised by additional expensive and complex procedures like bending, stamping, drilling etc. Also the inserted material has to be fabricated and placed into the mould, increasing the cost for the automation and the supply of the semi-finished materials.

Summarised, the various processes have disadvantages resulting from limitations in productivity, processing properties or the level of achievable geometrical part complexity.

Within the scope of the Cluster of Excellence "Integrative Production Technology for High-Wage Countries", two new approaches are developed to overcome the described disadvantages. The injection moulding of plastics and pressure die-casting of metal on one side, and the injection moulding and thermal spraying of metal on the other side are two new integrated processes for the production of plastic/metal hybrid parts.

10.2 Integrated Metal/Plastics Injection Moulding (IMKS)

The Integrated Metal/Plastics Injection Moulding (IMKS) is constantly developed since 2007. The IMKS uses the established and proven method of the multi-component technology as the fundament to combine the plastics injection moulding with the metal pressure die-casting to one integrated process (Fig. 10.2). With the IMKS it is possible to on-mould conductive tracks on a primarily injection moulded plastics carrier. The alloys used on the basis of tin have high electrical conductivities and are already established in the field of lead-free soldering for electronic applications. In addition, the melt temperatures of these alloys, within a range between 200–250 °C, fit into the temperature range of the used engineering thermoplastics (e.g. PA6.6 or PBT). The essential development cores for the technical implementation of the IMKS are the appropriate choice of materials, the development of an injection technology for reproducible processing of the low-melting metal alloys and a mould technology to produce ready-to-use electronic parts with integrated functions.

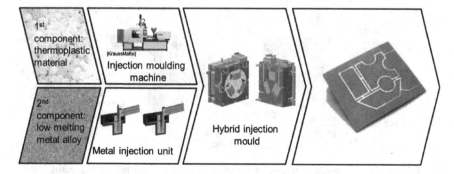

Fig. 10.2 Process chain "Integrated Metal/Plastics Injection Moulding"

10.2.1 Device for the Processing of Low-Melting Metal Alloys

While accessory units as an extension of standard machines to multi-component injection moulding machines are state of the art since the early 1990s, there is no similar solution in the field of metal pressure die-casting. For this reason, based on studies at the IKV and supported by the Krallmann Plastics Processing Gmbh, Hiddenhausen, Germany, a compact accessory unit for processing of low melting metal alloys has been developed. It can be integrated into the injection moulding process via core pull control (Hopmann et al. 2011, 2012, 2013). Based on the hot chamber die casting, the accessory unit utilises a plunger which is completely immersed in the liquid metal. The metal alloy is supplied in commercial bar form in the open, electrically heated melting pot and melted by heat conduction. The dosing is done by retracting the injection plunger. In this case, an overflow hole is released, whereby molten metal flows into the injection cylinder by gravity. Figure 10.3 shows the operation of the metal die-casting unit schematically.

10.2.2 IMKS Mould Technology

The IKV supported by the Krallmann Plastics Processing GmbH developed a 3-station index plate mould which allows the processing of two plastics and one low melting metal alloy in one mould and one machine (Fig. 10.4).

The molten metal and two different plastic melts are supplied to the respective cavities via hot runner valve gate nozzles. The transfer between the individual stations is done via a servo-electric powered index plate. The index plate technology has proved to be the most appropriate mould technology to produce complex multi-component metal/plastic components due to their geometrical freedom on the closing and nozzle side.

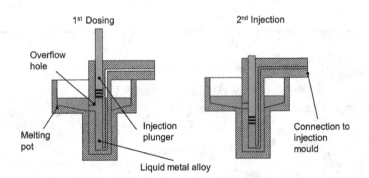

Fig. 10.3 Dosing and injection process of the metal die casting unit for the Integrated Metal/Plastics Injection Moulding (IMKS)

Injection
plunger

Cavities 1-3

Melting
pot

Fig. 10.4 3-station index plate injection mould with side-mounted metal die casting unit

By using this mould technology, as presented on the K-show 2010 in Düsseldorf, Germany, the fully automated production of a three component sports glasses with integrated conductive tracks for the heating and defogging of the lenses has been realised (Michaeli et al. 2010). The new process offers high reproducibility and short cycle times which qualifies the new technique for industrial production. The glasses were moulded within a cycle time of 80 s. Thereby a three dimensional conductive track with varying cross-sectional area was manufactured featuring high aspect ratios and the possibility of a direct contacting of metallic inserts. Additionally presented on the Fakuma-show 2012 in Friedrichshafen, Germany, a second demonstrator in the form of a desk lamp shows the possibility of the direct in-mould solder like connecting the conductive tracks with a LED (Doe 2012). The LED is already integrated into the part in the process. Thus no assembly is required at all to produce the lamp (Fig. 10.5).

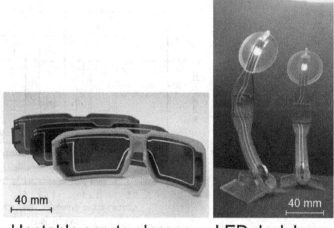

40 mm

40 mm

Heatable sports glasses   LED desk lamp

Fig. 10.5 Demonstrators for the IMKS

10.2.3 Influence of Variothermal Mould Temperature Control on the Achievable Conductive Track Length

In order to provide complex conductor path structures while minimising the material consumption of the metal alloy, conductor paths as filigree as possible should be designed, for example with a diameter of less than 1.5 mm^2 and a length of several hundred millimeters. To achieve the flow length by using the IMKS the use of a variothermal mould temperature control is expected to be advantageous. By local and close-to-cavity heating of the mould in the area of the conductor paths prior to the metal alloy injection and rapid cooling after the injection, it is possible to produce such conductor paths and other filigree structures without thermally damaging the plastics carrier.

For the investigation of the flowability of the metal alloy a meandering flow channel course is milled into a polyamide 6 carrier plate using a CNC-driven milling machine and different cross-sectional dimensions (Fig. 10.6). The carrier plate has a thickness of 4 mm. The flow channel is filled with the low melting metal alloy in an experimental mould.

To achieve a variothermal process control an inductor is moved into the open mould via a 6-axis robot of the KUKA AG, Augsburg, Germany. Subsequently, the surface of the nozzle half is heated in the region of the conductor paths by the inductive alternating field, leading to a temperature above the melting point of the metal alloy of approximately 230 °C during the injection phase. The carrier plate is then manually inserted into the mould cavity on the closing half, the inductor is swivelled out from the mould, the mould closes and the metal alloy is injected.

An increase of the flow length of the metal component due to the variothermal mould heating can be observed for all cross-sectional dimensions (Fig. 10.6, right). The high mould wall temperature leads to a delayed solidification, resulting to a

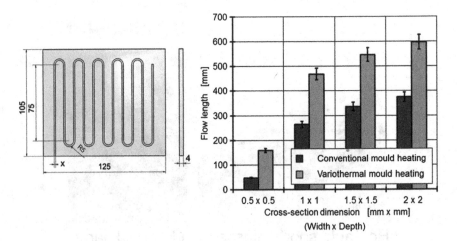

Fig. 10.6 Test specimen and results of the flow length investigations

doubling of the achievable flow length especially for cross-sections below 1×1 mm^2. It is also clear that the results of the samples prepared with variothermal mould temperature control have a larger scatter. The reason for this is not yet fully understood, but subject to further investigation currently ongoing. Since the components are inserted manually, the slightly different retention time of the plastics carrier plates in the mould can be a source of variation.

10.3 In-Mould-Metal-Spraying (IMMS)

The In-Mould-Metal-Spraying (IMMS) is another integrated process designed to simplify the production of metallised plastics components. At the same time it opens up new possibilities in product development. The IMMS enables the fabrication of metallic coatings onto the surface of plastics components, which can be used in the electrical industry. Certain requirements on these components regarding the haptic characteristics (cool-touch-effect) and electromagnetic shielding (EMC) can be fulfilled with metal coatings. With this method, under development since 2012 by the IKV and IOT at the RWTH Aachen University, at first a metal coating is applied inline to certain areas of the cavity surface of an injection mould via thermal spraying. In the next step the metal layer is back-moulded with plastic. The metal layer is thereby transplanted to the plastics component, similar to the in-mould labelling process. Metal layer and plastic are then demoulded as a plastic component with integrated partially metallised surface (Fig. 10.7).

10.3.1 Selection of Materials and Thermal Spraying Process

The thermal spraying process has to fulfil special requirements to enable the integrated process. One key question is, if the thermal sprayed coating can be separated

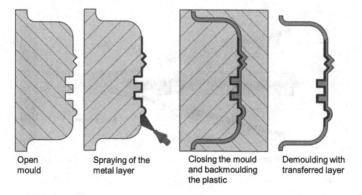

| Open mould | Spraying of the metal layer | Closing the mould and backmoulding the plastic | Demoulding with transferred layer |

Fig. 10.7 Schematic of the In-Mould-Metal-Spraying (IMMS)

from the mould surface during injection moulding. Therefore, in particular the adhesive strength between metal coating and the overmoulded plastics has to exceed the adhesive strength between metal coating and mould surface. Simultaneously the adhesion of the metal coating to the mould surface has to be sufficiently high to withstand the emerging shear stress during the injection moulding process. Also the surface of the mould shall not be destroyed through the thermal spraying process to enable a continuous reproducible production.

Subject to these conditions the wire arc spraying process is chosen for the tests (Fig. 10.8).

The particle velocities in wire arc spraying process are, in comparison to the other conventional thermal spraying processes relatively low (50–100 m·s^{-1}). Thereby, a low abrasion of the mould surface is expected. Another conventional thermal spraying process, which exhibits similar particle velocities, flame spraying, cannot reach deposition rates reached by wire arc spraying. The cooler deposit characteristic of wire arc spraying minimizes the substrate heating common with other thermal spray processes; hence processes advantages regarding the substrate temperatures (Tucker 2013). Zinc is chosen to be the deposition material for the metallic component. Zinc coatings are widely used for corrosion protection, often for steel substrates. Moreover the compatibility to different plastic components is given in earlier experiments where the zinc coating was brought onto the plastic surface, using the wire arc spraying (Bobzin et al. 2011).

Without a special surface treatment of the used cavity insert it was possible to transfer >95 % of the zinc coating onto the plastics part (Fig. 10.9). The milling marks provide enough adhesion.

Through application of masking upon the mould surface before the thermal spraying process, partial transplantation of the selected areas onto the plastics component can be realised. This extends the degree of freedom during the metallisation process. As mentioned before, the adhesion of the metal coating to the mould surface must be carefully adjusted to fulfil the requirements regarding the transferability of the coatings in the IMMS process. A key factor hereby is the

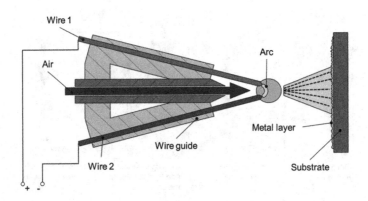

Fig. 10.8 Schematic of the wire arc spraying process

Fig. 10.9 Transplantation of complex geometries

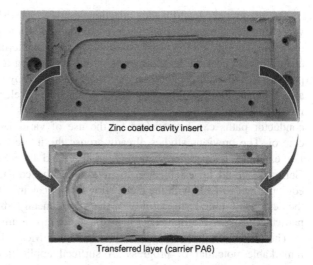

Zinc coated cavity insert

Transferred layer (carrier PA6)

Fig. 10.10 Selective transplantation with IMMS

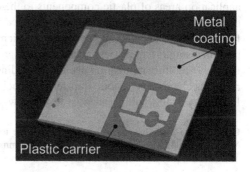

roughness of the mould surface. In order to achieve the intact transplantation of the coatings, different roughening methods and parameters have been investigated and analysed (Bobzin et al. 2014). As a result, together with the usage of masking methods, selective transplantation of the metal coating was realised. The transplanted coating was intact (Fig. 10.10).

10.4 Conclusion and Outlook

The presented process combinations can provide benefits in terms of a more efficient production of plastic components with electrically conductive elements. Both methods, developed at the Cluster of Excellence "Integrative Production Technology for High-Wage Countries" at RWTH Aachen University, increase the added value of electrical products, at the same time rationalise the production and thus help companies of high-wage countries to remain competitive in the future.

The Integrated Metal/Plastics Injection Moulding (IMKS), evolved over the last years, enables the production of electrical and electronic parts in extremely short cycle times compared to existing multi-stage process alternatives. The short process chain is made possible by the transfer of the underlying technologies in a one-step process, in which all the necessary steps are controlled by one machine. The studies also indicate the possibility of direct contacting of electrical inserts inside the mould, which in turn saves additional process steps. The demand for filigree conductor paths can be satisfied by the use of variothermal mould temperature control. The previous studies already arouse the interest of industrial companies. For example, cooperation between the IKV and the OSRAM GmbH, Munich, Germany, could be established to investigate the contacting of a circuit board equipped with a LED with the metal alloy for optical applications. In future studies, the achievable long-term stability of the plastic/metal hybrids will be investigated particularly under the influence of media and temperature loads.

The studies on the newly developed In-Mould-Metal-Spraying (IMMS) show a remarkable potential of progress. A surficial application of the metal coating through the combination of wire arc spraying and injection moulding can extend the application areas of plastic components to the electronic industry by utilizing these components with better electromagnetic shielding and haptic characteristics. In further studies, a new mould technology which enables the production of the IMMS parts in a single-step process shall be investigated. Well-known mould technologies from the multi-component injection moulding like e.g. transfer processes using rotary mechanism or robotic transfer (Michaeli and Johannaber 2004; Michaeli and Lettowsky 2005) will be adapted to the new process.

Acknowledgments The depicted research referring to the Integrated Metal/Plastics Injection Moulding (IMKS) and the In-Mould-Metal-Spraying (IMMS) has been funded by the German Research Foundation (DFG) as part of the program Cluster of Excellence "Integrative Production Technology for High-Wage Countries" at the RWTH Aachen.

We would like to extend our thanks to the DFG, the German Bundesministerium für Wirtschaft und Energie (BMWi). We also thank all companies who have supported these research projects through the provision of materials, machinery and other resources.

References

Berneck J (2011) Kunststoff statt Metall. Kunststoffe 102 (9):109–111
Bobzin K, Michaeli W, Brecher C, Kutschmann P (2011) Integrative Produktionstechnik für Hochlohnländer. Springer-Verlag, Heidelberg
Bobzin K, Öte M, Linke TF, Schulz C, Hopmann C, Wunderle J (2014) Integration of Electrical Functionality by Transplantation of Cold Sprayed Electrical Conductive Cu Tracks via Injection Moulding. In: International Thermal Spray Conference 2014, Barcelona, 2014. DVS-Berichte, vol 302

Brosig E (1996) Chemisch-Kupfer sorgt für eine schützende Haut. EMV-Schutz: Kunststoffgehäuse selektiv metallisieren. Industrieanzeiger 43:44–46

Chanda M, Roy K (2007) Plastics Technology Handbook. CRC Press, Taylor & Francis Group, Boca Raton

Doe J (2012) Erfolgreiche Fakuma für KraussMaffei, Netstal und KrausMaffei Berstorff. München

Drummer D, Dörfler R (2007) Mechatronik mit Kunststoffen–Herausforderungen auf dem Weg vom Werkstoff zur Baugruppe. Paper presented at the Spritzgießen 2007, Düsseldorf

Flepp A (2012) Wirschaftlicher als Metall. Kunststoffe 102 (8):73–75

Grob W, Müller K, Habiger E (2003) EMC Kompendium 2003. publish-industry, München

Hopmann C, Neuß A, Wunderle J (2011) Hybrid multi-component injection moulding for electro- and electronic applications. Paper presented at the Proceedings of the 27th World Congress of the Polymer Processing Society, Marrakesh, Marokko

Hopmann C, Neuß A, Wunderle J (2012) Integrierte Fertigung von E&E-Bauteilen durch hybrides Mehrkomponenten-Spritzgießen. In: Umdruck zur VDI-Jahrestagung Spritzgießen Baden-Baden, 2012

Hopmann C, Neuß A, Wunderle J (2013) Fertigung von komplexen Elektronikkomponenten durch Integriertes Metall/Kunststoff-Spritzgießen (IMKS). Paper presented at the VDI-Fachtagung Polytronics, Frankfurt am Main

Kanani N (2009) Galvanotechnik: Grundlagen: Verfahren und Praxis einer Schlüsseltechnologie. Carl Hanser Verlag, München, Wien

Michaeli W, Grönlund O, Neuss A, Wunderle J, Gründler M (2010) New Process for Plastic Metal Hybrids. Kunststoffe International 9 (2):102–105

Michaeli W, Johannaber F (2004) Handbuch Spritzgießen. Carl Hanser Verlag, München, Wien

Michaeli W, Lettowsky C (2005) Mehrkomponentenspritzgießen. Verfahren und Möglichkeiten. In: Umdruck zur VDI-Fachtagung Spritzgießen, Baden-Baden, 2005

Pfeiffer B Elektrisch leitfähige Kunststoffe. In: OTTI Technik-Kolleg, Regensburg, 2005

Pflug G (2005) Kunststoffgehäuse abschirmen. Kunststoffe 95 (2):22–27

Tucker RC (2013) Thermal Spray Technology, vol 5A. ASM Handbook. ASM International, Materials Park

Part V
Self-Optimising Production Systems

Christopher M. Schlick, Fritz Klocke, Barbara Deml, Dirk Abel,
Christian Hopmann, Thomas Auerbach, Jennifer Bützler, Marco Faber,
Stefan Graichen, Gunnar Keitzel, Sinem Kuz, Matthias Reiter,
Axel Reßmann, Thorsten Stein, Sebastian Stemmler, Drazen Veselovac

Today many production systems are highly automated, enabling premium quality and cost-effective high-volume manufacturing. Innovative automation technologies also make it possible to manufacture products in reliable non-stop operation, thereby greatly improving productivity in high-wage countries. However, increasing global competition and the resulting cost pressure require a more flexible adaptation of highly automated production systems to market conditions in order to meet the demand for variety and short product life cycles. Wiendahl et al. (2007) describes this phenomenon as the replacement of the era of mass production by the era of market niches. As a result, the product range increases due to multiple variants of the same product and a growth of the different types of products. In order to stay ahead of the competition in this turbulent environment, it is crucial for companies in high-wage countries to anticipate customer-specific wishes for an individual adaption of high-quality products and to react extremely flexibly. Against this background, ultra-flexible forms of production systems are required to continuously adapt to changing product structures and the corresponding production processes (Brecher 2012). These requirements are extremely difficult to meet with present automation technology as the necessary ultra-flexibility of function and behaviour cannot be achieved with conventional programmable subsystems (sensu Brecher 2012). Demand is therefore high for new concepts of automated planning, programming and control.

One approach to achieve a whole new level of flexibility is to design self-optimizing production systems that are capable of taking goal-oriented and task-focused action. Based on an active sensing and monitoring of the environment, these systems possess the ability to adjust their structure, function and behaviour as well as their internal goals autonomously according to perceived changes of the situation and the predicted consequences (Adelt et al. 2009). The internal goals usually represent high-level goals such as desired quality, lead time, throughput and utilization, whilst the task structure defines the goal-driven recursive decomposition of the major manufacturing task into subtasks. These self-optimising systems can range from single machines and manufacturing cells up to the factory level resulting in a cascade control scheme that can be modelled by a self-similar architecture.

Self-optimisation on cell level is described in the first paper "A symbolic approach to self-optimisation in production system analysis and control". This paper presents a cognitive control unit for an ultra-flexible robotic assembly cell. The cognitive control unit simulates human knowledge-based behaviour and is suitable for assembling products consisting of cubic parts of arbitrary structure under arbitrary part supply. As the skilled human operator is an integral part of this assembly cell, the working conditions in the human–robot cooperation are improved by using a graph-based planner which is able to reduce occupational safety risks and avoid dangerous work procedures. This human-oriented approach of self-optimisation is presented and embedded in a hierarchical architecture for self-optimising production systems.

The presented architecture can also be used to describe self-optimisation on a lower level. Here, one challenge is to master processes which lack deterministic control functions. The second paper entitled "Approaches of self-optimizing systems in manufacturing" describes two approaches of self-optimisation on machine level. The first approach demonstrates the autonomous generation of technology models as process knowledge is a key factor of production and is an integral part of a self-optimising system. First, the general self-optimisation approach for metal cutting processes as well as the process independent modelling methodology is presented. An innovative approach for milling processes is then introduced which includes the new concept for the autonomous generation of process knowledge. The second approach presented in this paper describes model-based self-optimised injection moulding that enables the compensation of process fluctuations in order to guarantee a constant part quality.

It is very unlikely that self-optimising technical systems will reach the flexibility and the knowledge, skills and abilities of a human being in the near future, so human work has to be regarded as an essential part of the production process. For example, highly developed sensorimotor skills are needed for bimanual handling and assembly of limp parts which are extremely difficult to automate. In order to enable an ergonomic cooperation between human and machine in manufacturing environments, the individual capabilities and limitations have to be considered. The third paper "Adaptive workplace design on the basis of biomechanical stress curves" describes an approach to perform physiological stress-oriented self-optimisation processes. By using motion capturing data of manual assembly tasks and a biomechanical body model, a functional description of stress was derived in terms of body-part-oriented stress curves for the upper extremities that enable an evaluation of the movements and handling positions. This stress-related assistant system can be used for adaptive automation to provide employee-specific support, e.g. in the supply of components. As a result, work systems will be enabled to optimise and adapt themselves to the individual abilities of the employees.

References

Adelt P, Donoth J, Gausemeier J et al (2009) Selbstoptimierende Systeme des Maschinenbaus – Definitionen, Anwendungen, Konzepte. In: Gausemeier J, Raming FJ, Schäfer W (ed) HNI-Verlagsreihe, Volume 234. Heinz Nixdorf Institut, Paderborn

Brecher C (2012) Integrative Production Technology for High-Wage Countries. Springer, Heidelberg

Wiendahl H, ElMaraghy H, Nyhuis P, Zäh M, Duffie N, Brieke M (2007) Changeable manufacturing – classification, design and operation. CIRP Ann Manuf Technol 56(2): 783–809

Chapter 11
A Symbolic Approach to Self-optimisation in Production System Analysis and Control

Christopher M. Schlick, Marco Faber, Sinem Kuz
and Jennifer Bützler

11.1 Introduction

With steadily increasing customer requirements on quality of both products and processes, companies are faced with increasing organisational and technical challenges. The market is characterised by individualised customer wishes which result in individual adaptations of the products. In order to manage this rapidly growing variety of products, the production system has to become much more flexible with respect to the product structure to be manufactured and the corresponding production and assembly processes. Especially in the field of assembly systems the increasing variety of products adds new complexities to the planning process and increases the costs, because (re-)planning efforts tend to grow exponentially to the number of variants.

One approach to overcome these limitations is to design production systems that are able to autonomously adjust to market needs. If the automatic control systems of machines, robots and technical processes could flexibly adjust themselves to the environmental conditions and autonomously find solutions through a goal-oriented forward and backward chaining of production rules, the efforts of developing the control programmes would be reduced significantly. This would cut down the non-value-adding activities, thereby yielding a higher productivity for the company. Following the seminal work of Adelt et al. (2009) we speak of self-optimisation.

Besides flexibility, companies also have to integrate the working person into the production process. The human operator will always be involved either by directly taking over assembly tasks (e.g. for limp components) or by supervising the assembly process. Furthermore, unique human skills such as sensorimotor coordination and creative problem solving cannot be automated. To establish a safe,

C.M. Schlick (✉) · M. Faber · S. Kuz · J. Bützler
Institute of Industrial Engineering and Ergonomics (IAW), RWTH Aachen University,
Bergdriesch 27, 52052 Aachen, Germany
e-mail: c.schlick@iaw.rwth-aachen.de

© The Author(s) 2015
C. Brecher (ed.), *Advances in Production Technology*,
Lecture Notes in Production Engineering, DOI 10.1007/978-3-319-12304-2_11

effective and efficient integration of the working person into the production process, ergonomic aspects have to be considered. New technologies such as lightweight robots or electro-optical sensors open up new possibilities in the area of ergonomic human-robot cooperation. For the first time, it is now possible to abolish the strict separation between the work areas of the human and the robot (e.g. Bascetta et al. 2011; Fryman and Matthias 2012; Matthias et al. 2011). Light detection and ranging sensors in particular enable the robot to recognise the human early enough to adjust or even stop its movement. The action forces of lightweight robots are also considerably lower than those of conventional industry robots, minimising the risk of injury and ensuring the safety of the cooperating working person.

In this regard a cognitive control unit has been developed that can cognitively control a robotic assembly cell. It is embedded into a general architecture for self-optimising production systems.

11.2 Cognitive Automation

In order to cope with the cited challenges for assembly systems a novel approach to cognitive automation was developed (Mayer 2012; Faber et al. 2013). To support the human operator effectively and efficiently, he/she has to be able to understand the system's functions and behaviour. A simplified compatible representation of the mental model of the operator on assembly processes in a dynamic production environment based on production rules has therefore been developed and integrated into the knowledge base of the cognitively automated system. By explicitly considering ergonomic criteria (e.g. feasibility, occupational risks, freedom of impairment, promotion of personality development (Luczak and Volpert 1987) the system is also capable of improving the working conditions for the human operator interacting, for instance, with the robot or supervising its functions. Figure 11.1 depicts the architecture of the cognitively automated system. The central element is the Cognitive Control Unit (CCU) which is based on the three layer architecture for robotic applications consisting of a planning, a coordination and a reactive layer according to Russel and Norvig (2003). The architecture has been extended with a presentation layer for ergonomic human-machine interaction and a technical layer that includes the sensors, automatic control algorithms and actuators (Hauk et al. 2008).

11.2.1 Cognitive Automation of Assembly Tasks

The cognitive automation functions for self-optimising assembly processes are realised in the planning layer, the central element of the CCU. This layer is responsible for planning and optimising the assembly sequence and for deriving high-level action commands according to the generated assembly steps. In contrast, the reactive layer is responsible for the direct communication with the actuators and

Fig. 11.1 Architecture of the
Cognitive Control Unit
(adapted from Mayer (2012))

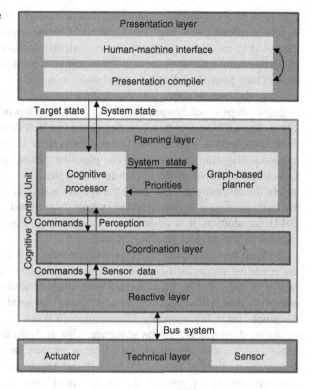

sensors. The coordination layer in between translates between the planning and
reactive layer. A detailed description of all three layers can be found, for example,
in Hauk et al. (2008), Mayer et al. (2012) and Faber et al. (2013). The following
section will focus on the planning layer. To evaluate the functions of the CCU a
cognitively automated assembly cell has been developed (Brecher et al. 2012).
A six axes articulated robot (KUKA KR30 Jet) is used with a three finger gripper
with haptic sensors (SCHUNK SDH2) to handle parts and components. The work
area of the assembly cell is divided into three sub-areas: Parts and components are
fed into the system through a circular conveyor belt. In addition, two sub-areas are
used to assemble the final product and to buffer parts and components that cannot
be assembled directly. The parts and components on the conveyor belt do not have
to be in a predefined sequence. The sequence can be completely random and may
also include parts that are not needed for the current product being assembled.

The final product is specified by the human operator through the human-machine
interface in the presentation layer and represents the goal state of the cognitive
controller. The goal state contains only the geometric information about the final
product including the type, position and orientation of the individual components in
terms of CAD data. This data is forwarded in combination with the planning
knowledge to the cognitive controller. Based on the goal state and the current
system state that is propagated by the coordination layer, the cognitive processor is

able to derive the assembly sequence autonomously. The optimal next assembly step is transferred as a high-level command to the coordination layer where it is translated to machine commands for the articulated robot used.

The decision-making process of the cognitive processor is based on the cognitive architecture Soar (Laird 2012), a symbolic computational system that is able to simulate the human cognition. The knowledge that is necessary for planning the assembly steps is solely specified in terms of if-then production rules (Faber et al. 2014). The CCU is able to adjust flexibly to changes in the part sequence, because there is no need to (re-)estimate parameters as there is with other methods such as dynamic Bayesian networks. The planning knowledge includes procedural knowledge of experienced operators and is therefore represented in a way that makes the assembly process more transparent and conforms to the expectations of the human operator supervising the system (Mayer and Schlick 2012; Faber et al. 2014). In addition, it is designed as generically as possible so that it can handle changes in the product structure as well.

To keep the complexity of the production rules and the planning process within the cognitive processor low, the processor has a very limited planning depth. In fact, it is only able to plan one assembly step in advance. However, this is not enough to deal with complex planning criteria that need to take information about the whole assembly sequence into account in order to ensure that safety-critical situations do not occur in the sequence. This is essential for ergonomic working conditions, because the safety of the human operator has to be ensured at all times during the production process. To satisfy this requirement the cognitive processor was extended by a graph-based planner. This planner is described in detail in the next section. In this way, the ergonomic risk can be minimised and, if some risk is unavoidable, reduced to an acceptable level. In this case a warning message could be given to the human operator at specific points in time to alert him/her to the types and sources of risks.

11.2.2 Adaptive Planning for Human-Robot Interaction

As mentioned in the previous section, the originally developed cognitive processor is purely reactive and is not able to consider complex optimisation criteria in the planning process. Extending its planning process to a higher planning depth (or even to a full planning process considering the complete assembly sequence) would inevitably result in a much more complex planning procedure. An exponentially growing number of achievable goal states have to be simulated and compared against each other in order to find the optimal alternative. To make the cognitive processor more efficient it has been extended by a graph-based planner (Faber et al. 2014) whose operation mode follows a hybrid planning approach including an offline and an online phase (Ewert et al. 2012). It interacts with the cognitive processor and provides additional information about the future assembly sequence for the decision phase of the cognitive processor.

In preparation of the assembly process, the structure of the product is transferred into a directed state graph including all valid assembly sequences. In particular, each state represents an achievable intermediate goal state in the assembly process. The intermediate states are identified by recursively decomposing the final product according to the "assembly by disassembly" strategy (see e.g. Thomas and Wahl 2001). Consequently, each edge of the resulting graph can be considered as a feasible assembly step which modifies the intermediate product state by adding exactly one part or component. The generation of the assembly graph can be done offline because, despite changes in the product structure, the same dependencies can be used in every cycle without losing the flexibility of the cognitive processor to react to changes in the assembly environment. Figure 11.2 shows an exemplary assembly graph of a simple product consisting of five cubic parts.

In order to be able to compare the alternatives for the next assembly step, each edge is weighted with a set of costs indicating how "costly" it is to perform the corresponding assembly step. In its simplest form, each feasible assembly step induces costs c_b representing the basic effort of the assembly action. In addition, rule-based planning knowledge can be formulated and applied to consider additional optimisation criteria. These rules can be activated individually and refer to the state transitions of the assembly graph. If the condition of a rule is satisfied, its costs are added to c_b yielding a set of costs per edge depending on the activated planning criteria. Figure 11.2 demonstrates a simple scenario where a two finger gripper is

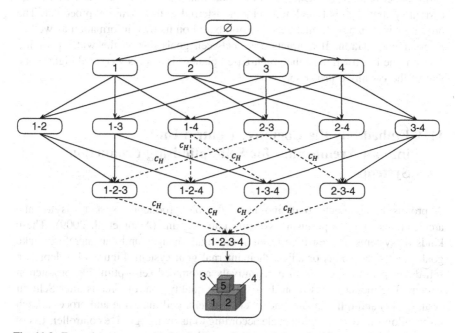

Fig. 11.2 Exemplary state graph of a simple product consisting of cubic parts. The *dotted edges* indicate assembly steps that have to be carried out by the human operator due to the technical restrictions of a two finger gripper (Faber et al. 2014)

used by an articulated robot to assemble the product. As the gripper requires two freely accessible parallel sides, the gripper cannot handle all components. At some points in the assembly process the human operator has to take over assembly tasks to assemble the final product (indicated with costs c_H). However, as one of the main objectives of the CCU is to assemble the product as autonomously as possible and to let the human operator take over an assembly task only if necessary, the number and sequence of manual interventions should be optimised. These manual interventions should be chosen in a way that the operator can effectively and safely use and develop his/her skills in the assembly process and has a complete work process. As can be seen in the graph, the interventions by the operator cannot be avoided completely, but can be optimised by selecting an assembly sequence in advance that leads to low physiological costs due to few and grouped interventions. Therefore, the right decision already has to be made on the second level of the presented graph, at which the cognitive processor itself does not have enough information in order to reliably choose the optimal path.

To be able to provide sufficient information to the cognitive processor, the graph-based planner has to evaluate the costs of the remaining assembly sequence in each assembly cycle starting at the current system state. Therefore a modified version of the algorithm A*Prune (Liu and Ramakrishnan 2001) is applied to the graph. The modifications refer to the modality of comparing two alternative assembly sequences in order to adjust the algorithm to the given application scenario (Faber et al. 2014). Once a set of k potential assembly steps fitting best to the current system state is found, this set is transferred to the cognitive processor. The processor is then able to make its decision based on its own information as well as external information. If conflicts arise between goals due to the wider planning horizon, the information of the graph-based planner is always weighted higher than that of the cognitive processor.

11.3 Embedding the Cognitive Control Unit into an Architecture for Self-optimising Production Systems

A promising approach to design more flexible production systems is to take architectures of self-optimising systems into account (Adelt et al. 2009). These kinds of systems are sensitive to environmental changes and can therefore make goal-oriented decisions or adjust their internal goal system. Figure 11.3 depicts a self-developed architecture of a cognitively automated self-optimising production system. The model is based on the cascading quality control circuits after Schmitt et al. (2012) and differentiates the levels segment, cell, machine and process. Each layer follows its own decision cycle according to its own cognitive controller. Every subordinated layer can be considered as a cognitively controlled system of the next higher level. The resulting cascade control leads to a self-similar structure of the

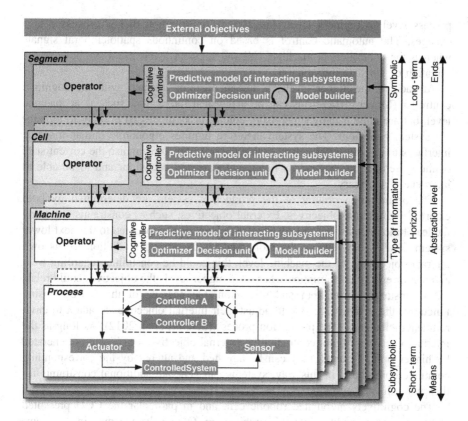

Fig. 11.3 Architecture for cognitively automated self-optimising production systems (adapted from Mayer (2012))

overall architecture that is comparable to hierarchically controlled software systems (e.g. Litoiu et al. 2005).

The bottom level of the architecture represents the sub-symbolic information processing of the automatic control systems. In the next higher levels, the adaptation process is based on symbolic "cognitive controllers". Their decision-making process is based on the current system state in conjunction with the pursued goal. In particular, they generate and update a model of the controlled process in conjunction with the environment within the model builder. This model contains the execution conditions of the production process as well as the information of the interacting subsystems in the appropriate granularity. Based on the generated model, the optimiser and decision unit are able to make context-sensitive decisions. At the machine level, for instance, functionalities of a model-based self-optimisation (Schmidt et al. 2012) are realised whereas the cell level aggregates several machines to higher level production units following coordinated actions. Finally, the segment can be considered as a macro structure combining several cells for the overall production process. The level of abstraction correspondingly increases from process level to segment level. The type of information that is processed also changes. The automatic control is based on continuous spatiotemporal signals whereas the controllers at machine, cell and segment levels use a symbolic representation of the state information.

At each of the higher levels, a human operator interacts with the cognitive controller (Mayer 2012). This can be a physical interaction, such as at machine level, but are more usually supervisory control tasks processed in order to monitor the system behaviour. The system therefore requires ergonomic human-machine interfaces to display information, enable the operator to recognise the current state of the system, to understand its functional state and behaviour and to be able to intervene if necessary.

The optimisation criteria of the production system are determined by both external and internal objectives. External objectives, such as constraints regarding the lead time or costs, are processed at each level and propagated to the next lower system. Each subsystem on the individual levels generates additionally its own internal objectives. At the machine level, this could be constraints regarding wear and tear or energy consumption whereas at higher levels the objectives could relate to, for instance, throughput and utilisation. On account of the self-optimising functions, the systems are able to adjust their internal objectives to adapt to environmental changes in the production process (Schmitt et al. 2012). As long as the internal objectives do not contradict the external objectives or objectives generated by higher order systems, they can be adjusted and altered by the corresponding cognitive controllers. In this way, systems can generate additional constraints for their subordinated systems.

The cognitively automated robotic cell, and in particular the CCU presented above, can be embedded into this architecture. Obviously, the machine elements such as the robot and the conveyor belt are located at the machine level. With the self-developed cognitive controller, the robot is capable of managing the pick and place process of individual parts and components in line with its own internal objectives. As the CCU focuses on automating the whole assembly cell, it is located at the cell level. The interacting subsystems of the cognitive controller are accordingly the assembly robot, the conveyor belt and the work areas. The main external objective is the assembly of the final product with respect to the given constraints (e.g. the part supply). To achieve this goal, a predictive model is used that contains the description of the final product in terms of CAD data and the knowledge about assembling the product. The knowledge comprises the production rules of the cognitive processor and the planning rules of the graph-based planner and forms the basis for the joint decision-making process.

The predictive model of the interacting subsystems is built by the model builder. The knowledge base required for the assembly process is formulated manually by production experts. This task has to be done with care as the knowledge affects not only the assembly process itself but also safety aspects of the human-robot interaction. Introducing erroneous production rules can lead to wrong decisions and non-acceptable risks for the human operator. In addition to the knowledge base, the internal representation of the final product is generated in the model builder for planning purposes. This representation also includes the automatically extracted neighbourhood relationships of the individual parts and components. Within the model builder it is also possible to generate new intermediate goals in order to divide the current task into smaller subtasks. Such subtasks could include managing the buffer area or removing erroneous components that have been misplaced (e.g. due to erroneous sensor readings).

Finally, the fusion of the data takes place in the optimiser and decision unit. In the optimiser the machine states are evaluated, including the available components. Based on the environmental model of the cognitive controller, the preferences in the material flow are set and alternatives in the action sequence are compared by the graph-based planner. The main goal of the optimiser is to reduce the solution space for the decision cycle of the cognitive software architecture Soar by providing action-oriented planning information. The decision for one of the possible actions is made in consideration of the preferences that have been set and the internal and external objectives of the subsystems involved.

11.4 System Validation

The function of the planning layer of the presented architecture has been validated by means of a simulation study. This study validated both the correctness of the generated assembly sequences and the support of human-robot interaction. Based on the developed architecture, the following hypotheses were formulated:

- The assembly process should be as autonomous as possible, so that the number of manual interventions within the human-robot cooperation is reduced to a minimum. Additionally, the type of manual tasks should let the human operator focus on his/her unique sensorimotor skills.
- To achieve a complete work process for the human operator, the manual work steps should be placed within the shortest possible time interval. The working person then has more flexibility in designing and organising his/her remaining work (supervisory control, quality control, etc.).
- If the product consists of several assembly groups, there should be as few changes between those groups during the assembly process as possible. This maximises the transparency of the assembly process and makes it easier to intervene if errors occur.

In a first simulation study, simple products consisting of single-type cubic parts were assembled (Faber et al. 2014). Both size and structure of the product were varied to yield assembly graphs of different complexity with respect to the average node degree. Products of type 1 consist of a single layer of parts whereas in products of type 5 all parts are mounted one above the other ("tower"). The structures in between describe intermediate complexities of the assembly graph. The part supply through the conveyor belt was completely randomized and included components that were not needed for the current product. The number of parts

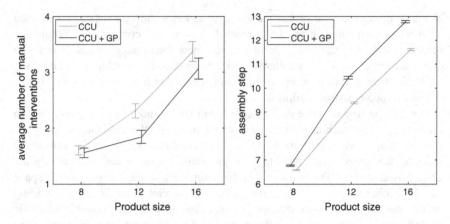

Fig. 11.4 Average number of manual interventions with activated and deactivated graph-based planner (*left*) and positions of the manual intervention in the assembly sequence of product type 2 (*right*)

that are concurrently fed into the system was also varied systematically. For each combination of the aforementioned independent variables, the assembly was simulated by the CCU with the graph-based planner either activated or deactivated. The cognitive planning of the assembly process had to be done under the following constraints: (1) New parts were only allowed to be assembled in the direct neighbourhood to existing parts in order to increase the transparency of the system behaviour (Mayer 2012). (2) The two finger gripper used needs two freely accessible parallel sides. Otherwise, this part has to be assembled manually by the human operator. Dependent variables for all simulation runs were the generated assembly sequence and the resulting number of manual interventions by the human operator.

Figure 11.4 shows the average number of assembly steps that have to be carried out by the human operator on the left side. Products of type 5 ("tower") are not considered here as they do not require human intervention. As shown in Fig. 11.4 (left) the manual interventions can be reduced for all product sizes. For products consisting of 12 parts this reduction is also significant ($p < 0.01$) according to a t-test with level of significance $\alpha = 0.05$. The distribution of the manual assembly steps in the assembly sequence could also be improved. On the right-hand side, Fig. 11.4 exemplarily shows the results for products of type 2. In this case, the interventions could be moved to a later point in time for product sizes larger or equal to 12 parts and almost fixed to a single point in time for products consisting of 8 or 16 parts.

To evaluate the third hypothesis and to transfer the approach to a real product, a second simulation study was carried out (Schlick et al. 2014). In this second study, a model of a Stromberg carburetor consisting of three independent assembly groups was assembled. A new planning rule was introduced in the graph-based planner prohibiting a new assembly group from being started while other assembly groups were still not finished. The simulation study covered three scenarios: (1) planning

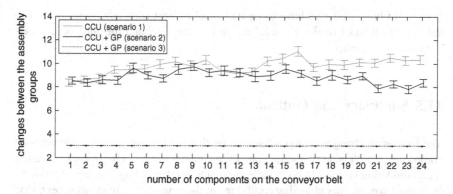

Fig. 11.5 Average number of changes between assembly groups of the Stromberg carburettor depending on the number of components that are fed into the system at the same time (adapted from Schlick et al. (2014))

without the graph-based planner, (2) planning with the graph-based planner, whereby it was allowed to ignore the cited planning rule and (3) planning with the graph-based planner, whereby the rule had to be obeyed. The part supply was again completely randomised. The number of supplied parts was varied systematically between 1 and 24. In all cases, the central part, on which the other parts are assembled, was supplied first.

The simulation results show that the new planning rule has an impact on the number of changes between the assembly groups (Fig. 11.5). Using the CCU with deactivated graph-based planner (scenario 1) yields an average number of changes of 9.78 ($SD = 0.65$). In scenario 2 the graph-based planner is activated and consequently the assembly of an assembly group should preferably be finished before starting a new one (but this is not obligatory). This effect can be reproduced by the simulation. It was possible to significantly reduce the number of changes between the assembly groups according to the Wilcoxon signed-rank test at a level of significance of $\alpha = 0.05$ ($mean = 8.86$, $SD = 0.53$, $p < 0.001$). In contrast to scenario 2, the third scenario requires one assembly group to be finished before starting a new one. In this case, it was always possible to reach the minimum number of three changes. In summary, the simulation study shows that using the graph-based planner significantly reduces the number of changes between the assembly groups and thereby improves the transparency of the system behaviour for the human operator.

However, the scenarios require different efforts for managing the component flow, because supplied components that are not allowed to be assembled directly have to be stored in a buffer. Consequently, more motion cycles (pick and place) are required yielding a higher assembly time for the product. In scenario 2 there was only an average increase of the pick and place operations of 0.84 % compared to scenario 1, whereas in scenario 3 the increase was 63.66 %. The reason behind this significant increase is the fixed rule of prohibiting alternating between assembly groups. So both scenarios have to be traded off against each other with respect to the improvement of working conditions on the one hand and the additional efforts required on the other.

11.5 Summary and Outlook

The increasing changeover to customised production imposes new requirements on companies, which want to remain competitive on the market. They have to redesign their production systems to be flexible enough to produce a huge variety of products in product space under changing conditions of the manufacturing environment. One approach to cope with this kind of complexity is to design self-optimising production systems according to a hierarchical system model. Each level can be considered as a self-optimising system in itself that controls the interacting subsystems. The cognitive controller on each level adjusts its predictive model accordingly and makes goal-oriented decisions on the basis of an optimiser and a decision unit.

The architecture was successfully validated by developing a cognitive control unit (CCU) for a robotic assembly cell. The CCU is able to cope with a large number of product variants, changes in the product structure and variability in the part supply. Its cognitive processor is based on the cognitive software architecture Soar. In order to be able to consider complex planning criteria such as ergonomic aspects, the cognitive processor is enhanced by a graph-based planner. It works on a dynamic state graph that contains all valid assembly sequences and whose edges are weighted according to the planning knowledge. Two simulation studies have shown that the CCU could successfully assemble products under completely randomised part supply and at the same time significantly improve the working conditions for the human operator. In future, the planning knowledge has to be enriched with further ergonomic knowledge in order to further improve human posture, movements and action forces in direct human-robot interaction.

The presented architecture could also be successfully applied to a sub-symbolic level of self-adaptive milling processes based on an adaptive model predictive control algorithm. First approaches concerning the prediction of parameters such as the dead time and the system matrix have produced promising results for future research.

Acknowledgments The authors would like to thank the German Research Foundation DFG for its kind support within the Cluster of Excellence "Integrative Production Technology for High-Wage Countries".

References

Adelt P, Donoth J, Gausemeier J et al (2009) Selbstoptimierende Systeme des Maschinenbaus – Definitionen, Anwendungen, Konzepte. In: Gausemeier J, Raming FJ, Schäfer W (ed) HNI-Verlagsreihe, Volume 234. Heinz Nixdorf Institut, Paderborn

Bascetta L, Ferretti G, Rocco P, Ardö H, Bruyininckx H, Demeester E & Di Lello E (2011) Towards safe human-robot interaction in robotic cells: an approach based on visual tracking and intention estimation. In: IEEE/RSJ International Conference on Intelligent Robots and Systems (IROS), p 2971–2978

Brecher C, Müller S, Faber M & Herfs W (2012) Design and Implementation of a Comprehensible Cognitive Assembly System. In: Conference Proceedings of the 4th International Conference on Applied Human Factors and Ergonomics (AHFE), USA Publishing, p 1253–1262

Ewert D, Mayer M Ph, Schilberg D & Jeschke S (2012) Adaptive assembly planning for a nondeterministic domain. In: Conference Proceedings of the 4th International Conference on Applied Human Factors and Ergonomics (AHFE), p 2720–2729

Faber M, Kuz S, Mayer M Ph & Schlick C M (2013) Design and Implementation of a Cognitive Simulation Model for Robotic Assembly Cells. In: Engineering Psychology and Cognitive Ergonomics. Understanding Human Cognition. Springer Berlin Heidelberg, p 205–214

Faber M, Petruck H, Kuz S, Bützler J, Mayer M Ph & Schlick, C M (2014) Flexible and Adaptive Planning for Human-Robot Interaction in Self-Optimizing Assembly Cells. In: Advances in The Ergonomics in Manufacturing: Managing the Enterprise of the Future, CRC Press, p 273–283

Frymann J & Matthias B (2012) Safety of Industrial Robots: From Conventional to Collaborative Applications. In: Proceedings of 7th German Conference on ROBOTIK 2012

Hauk E, Gramatke A & Henning K (2008) Cognitive Technical Systems in a Production Environment. In: Proceedings of the Fifth International Conference on Informatics in Control, Automation and Robotics. ICINCO: Madeira, Portugal

Laird J E (2012) The Soar Cognitive Architecture. MIT Press

Liu G & Ramakrishnan K (2001) A*Prune: an algorithm for finding K shortest paths subject to multiple constraints. In: Proceedings of the 20th Annual Joint Conference of the IEEE Computer and Communications Societies, p 743–749

Litoiu M, Woodside M, Zheng T (2005) Hierarchical Model-based Autonomic Control of Software Systems. In: Proceedings of the 2005 workshop on Design and evolution of autonomic application software (DEAS). St. Louis, Missouri, USA

Luczak H and Volpert W (1987) Arbeitswissenschaft. Kerndefinition – Gegenstandskatalog – Forschungsgebiete. RKW-Verlag, Eschborn

Matthias B, Kock S, Jerregard H, Kallman M, Lundberg I & Mellander R (2011) Safety of collaborative industrial robots: Certification possibilities for a collaborative assembly robot concept. In: IEEE International Symposium on Assembly and Manufacturing (ISAM)

Mayer M Ph (2012) Entwicklung eines kognitionsergonomischen Konzeptes und eines Simulationssystems für die robotergestützte Montage. Shaker Verlag, Aachen.

Mayer M Ph & Schlick C M (2012) Improving operator's conformity with expectations in a cognitively automated assembly cell using human heuristics. In: Conference Proceedings of the 4th International Conference on Applied Human Factors and Ergonomics (AHFE), USA Publishing, p 1263–1272

Mayer M Ph, Odenthal B, Ewert D et al. (2012) Self-optimising Assembly Systems Based on Cognitive Technologies. In: Brecher C (ed) Integrative Production Technology for High-Wage Countries. Springer, Berlin

Russel S J & Norvig P (2003) Artificial Intelligence: A Modern Approach. 2nd edition. Prentice Hall

Schlick C M, Faber M, Kuz S & Bützler J (2014) Erweiterung einer kognitiven Architektur zur Unterstützung der Mensch-Roboter-Kooperation in der Montage. In: Industrie 4.0 - Wie intelligente Vernetzung und kognitive Systeme unsere Arbeit verändern, Schriftenreihe der Hochschulgruppe für Arbeits- und Betriebsorganisation e.V. (HAB), p 239–263

Schmitt R, Brecher C, Corves B et al. (2012) Self-optimising Production Systems. In: Brecher C (ed) Integrative Production Technology for High-Wage Countries. Springer, Berlin

Thomas U & Wahl F (2001) A System for Automatic Planning, Evaluation and Execution of Assembly Sequences for Industrial Robots. In: Proceedings of International Conference on Intelligent Robots and Systems. p 1458–1464

Chapter 12
Approaches of Self-optimising Systems in Manufacturing

Fritz Klocke, Dirk Abel, Christian Hopmann, Thomas Auerbach, Gunnar Keitzel, Matthias Reiter, Axel Reßmann, Sebastian Stemmler and Drazen Veselovac

Abstract Within the Cluster of Excellence "Integrative Production Technology for High-Wage Countries" one major focus is the research and development of self-optimising systems for manufacturing processes. Self-optimising systems with their ability to analyse data, to model processes and to take decisions offer an approach to master processes without explicit control functions. After a brief introduction, two approaches of self-optimising strategies are presented. The first example demonstrates the autonomous generation of technology models for a milling operation. Process knowledge is a key factor in manufacturing and is also an integral part of the self-optimisation approach. In this context, process knowledge in a machine readable format is required in order to provide the self-optimising manufacturing systems a basis for decision making and optimisation strategies. The second example shows a model based self-optimised injection moulding manufacturing system. To compensate process fluctuations and guarantee a constant part quality the manufactured products, the self-optimising approach uses a model, which describes the pvT-behaviour and controls the injection process by a determination of the process optimised trajectory of temperature and pressure in the mould.

F. Klocke · T. Auerbach (✉) · G. Keitzel · D. Veselovac
Laboratory for Machine Tools and Production Engineering (WZL) of RWTH Aachen University, Steinbachstr. 19, 52074 Aachen, Germany
e-mail: t.auerbach@wzl.rwth-aachen.de

D. Abel · M. Reiter · S. Stemmler
Institute for Automatic Control (IRT), RWTH Aachen, Steinbachstraße 54, Aachen, Germany
e-mail: s.stemmler@irt.rwth-aachen.de

C. Hopmann · A. Reßmann
Institute of Plastics Processing (IKV) RWTH Aachen University, Pontstr. 55, 52052 Aachen, Germany
e-mail: zentrale@ikv.rwth-aachen.de

C. Brecher (ed.), *Advances in Production Technology*,
Lecture Notes in Production Engineering, DOI 10.1007/978-3-319-12304-2_12

12.1 Self-optimising Systems in Manufacturing

The industrial production is caught between uncertainties and relative lacksof precision (upper part of Fig. 12.1). Higher diversity of variants, smaller batch sizes, higher quality standards and increasing material diversities are conflicting priorities in the industrial production that have to be concerned in the future. The lower part of Fig. 12.1 illustrates the vision of process optimisation using sensor and control technologies to reduce variations in quality in contrast to conventional production without optimisation strategies. Self-optimising systems are high level control structures with abilities to analyse data, to model manufacturing processes and to make decisions where deterministic control functions do not exist.

A general overview on self-optimisation including a precise definition is given by Adelt et al. (2009). Approaches to integrate self-optimisation into technical processes and systems are manifold. Klaffert (2007) presents a self-optimising motor spindle that adjust its dynamic properties according to the respective machining situation autonomously. Kahl (2013) transferred the self-optimisation idea to the design process of mechatronic systems in order to improve the manageability of the complete development process.

To achieve the visionary scenario of production, research activities within the Cluster of Excellence focus on the development of self-optimising manufacturing systems. Therefore, a generic framework has been defined in the first development phase, compare Thombansen et al. (2012). In Fig. 12.2 the basic structure of the model-based self-optimisation approach and its modules are shown. The approach is structured in two parts: The "Model-based optimisation system (MO-System)" and the "Information processing Sensor and Actuator system (ISA-System)".

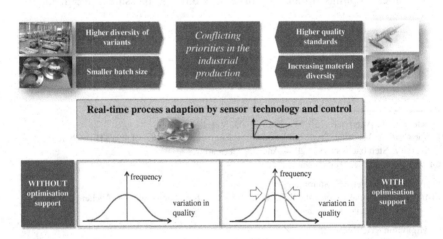

Fig. 12.1 Industrial production caught between uncertainties and relative lacks of precision—Klocke (2014)

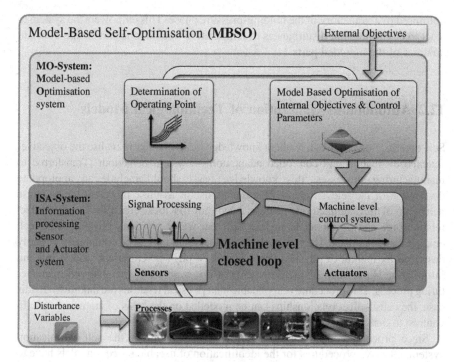

Fig. 12.2 The model based self-optimisation system—Thombansen et al. (2012)

The MO-System is the upper layer of the self-optimisation and implies the determination of optimal operating points and the self-optimisation strategies. The input parameters of the MO-system are the production plants external objectives; the output parameters of the MO-system are internal objectives and optimised control parameters for the ISA-system. The ISA-system is a real-time control loop with intelligent data analysis, sensors and actuators. The most challenging tasks for an implementation of the self-optimisation systems are on the one hand the identification of appropriate model-based optimisation strategies and on the other hand the provision of required data from the process provided by the used sensors. Most of the nowadays used sensor systems are not able to fulfil these requirements, as the data they provide are not directly usable as an input parameter for the above described system. Consequently, new sensor and monitoring systems have to be developed for the acquisition of real process data. Further challenges for establishing self-optimisation systems in production focuses also on social-technical aspects. It has to be addressed, how humans are able to interact with the self-optimising systems and how transparency at any state of the process can be ensured. In the following two chapters implementation examples are shown. The first example demonstrates the autonomous generation of technology models and the generation of technology knowledge, which is the core requirement of self-optimising systems. In the second example an established model of the pvT-behaviour in injection moulding is used to calculate the optimised pvT-trajectory of the holding-pressure phase. This empowers the system to react to environmental disturbances as temperature fluctuations and ensure constant qualities of the moulded parts.

12.2 Autonomous Generation of Technological Models

Self-optimisation requires a resilient knowledge basis in order to realise the objective-oriented evaluation and controlled adaptation of system behaviour. Transferred to manufacturing processes, this knowledge basis should include an appropriate description of the relevant cause-effect relationships as these represent the response behaviour of the manufacturing process. According to Klocke et al. (2012), cause-effect relationships can be modelled in four different ways: physical, physical-empirical, empirical and heuristic. The first two assume that relations can be completely or partly described by natural or physical laws. In case of physical-empirical models missing information is provided by measurements or observations of the analysed manufacturing process. This procedure is applicable if all physical relations are unknown. In this case, the cause-effect relationships can be modelled on the basis of empirical data. In contrast to that, heuristic models are derived from expert knowledge.

Since process models are an important prerequisite for the self-optimisation system, effective procedures for the identification of useable process models have to be developed. In this context, an innovative approach has been developed for the manufacturing process milling within the Cluster of Excellence. This development enables a standard machine tool to determine physical-empirical or empirical models for a given parameter space autonomously. This implemented system is illustrated in Fig. 12.3.

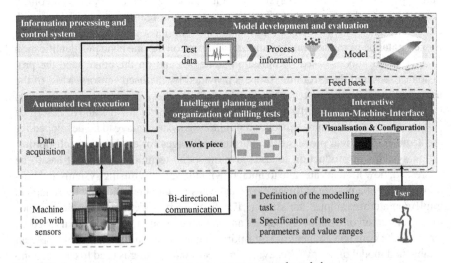

Fig. 12.3 Technology assistance system to generate process knowledge

Figure 12.3 shows the connection of an external information technology system (IT-system) to the machine tool. The IT-system fulfils two main functions. On the one hand, it operates as superior control system in order to realize the aspired system autonomy. On the other hand, the IT-system ensures the communication to the operator. Based on these two main functions, the following system modules have been designed and developed:

- An interactive human machine interface,
- a planning and organization procedure milling tests,
- an automated execution of milling tests and
- the automated modelling and evaluation of the conducted trials.

These system modules are described below.

12.2.1 Interactive Human Machine Interface

The communication to the operator is an important aspect. On the one hand, the autonomous system requires information of the used machine tool, the work piece, the cutting tool and the modelling task for its own configuration and documentation. Meta information on the test conditions are directly linked to the test results in order to enable a reuse of the obtained data and information. On the other hand, relevant system actions and the obtained test results need to be reported to the operator. Thus, a sufficient system transparency can be ensured, which ensures the acceptance of the autonomous system by the operator.

An interactive configuration wizard is developed for the first communication part. Interactive means in this context, that the input is checked for plausibility and the operator is alerted in case of incorrect entries. The technological limits of the machine tool and cutting tool are compared to the value ranges of the investigated parameters. Thus, it is not possible to define for example a cutting speed that will exceed the maximum spindle speed. Another example for the plausibility check is the comparison of entry data with technologically sensible limits. This supports the documentation process by identifying possible input errors such as a helix angle larger than 90°.

The second communication part is realised via a display window on an installed screen at the machine tool. This display is updated continuously while the autonomous system is running. It shows the planned test program, current actions like data transmission, test execution or model coefficient determination, as well as status messages such as "monitoring is active" or "disturbances occur". The illustrated information assists the operator to understand the behaviour and the decisions of the autonomous system.

12.2.2 Planning and Organisation of Milling Tests

As a first step, the planning and organisation module is responsible for the automated definition of test points. Test points are a suitable combination of feeds and speeds for a given test material. For this purpose, design-of-experiments methods are integrated into the autonomous system. Based on these methods the system determines appropriate parameter constellations which are investigated in milling tests.

When all test points are defined, the milling tests need to be distributed over the given work piece. This organisational step is required in order to define the starting positions of the tool during the automated testing phase. Figure 12.4 shows the approach to solve this distribution task.

Each milling test can be described as a rectangle with a certain width and height corresponding to the geometrical dimensions of the cut. Similarly, the lateral area of the work piece can be described by rectangular shapes. Based on this the so-called bin packing algorithms can be used to distribute the rectangles over a work piece, Dyckhoff (1990). On the upper right side of Fig. 12.4 an exemplary distribution result is illustrated. It shows a bin packing algorithm applied to rectangles which are pre-sorted according to their heights. Each of the rectangles and therewith the position of each milling test is thus clearly defined.

Before the planning and organisation phase can be completed the distribution result must be transferred to a machinable sequence of cuts which can be performed automatically. This includes not only the milling tests but also cuts which are needed to remove material and to clean the work piece. Cleaning cuts are necessary in order to avoid collision and to ensure accessibility to the next test cut. The determination of the whole cutting sequence is achieved by digitising the rectangles

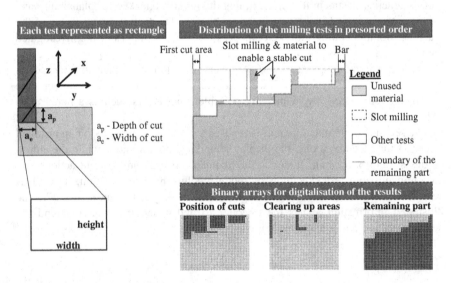

Fig. 12.4 Rectangle distribution

distribution. For that purpose, binary matrices with a defined grid size are used. The result of this process is also presented in Fig. 12.4.

12.2.3 Automated Execution of Milling Tests

The automation sequence uses a conventional line milling strategy for the execution of the milling trials. Because of this simple process kinematic the milling tests can be easily standardised and adapted to different cutting conditions. Furthermore, the starting and endpoint are clearly defined. This leads to a tool path, which can be easily implemented in a parameterised NC program.

Based on the standardised test procedure an automation sequence has been developed, which contains all steps such as the execution of milling operations, data acquisition as well as data analysis and processing. After each milling test the process relevant characteristic values are available and stored in a data base.

A further step focused on the implementation of an appropriate communication interface between the machine tool and the external IT-system. Via the communication interface several actions are realised. These are:

- **Triggering**: For a controlled process it is necessary to synchronise actions between machine tool and external IT-system. Trigger functions are used to announce that a sub system is ready.
- **Data transmission**: Values for process relevant parameter such as spindle speeds, feed velocities and tool centre point position need to be transferred from the external IT-system to the machine control. Therefore, a 16-bit data transmission has been installed.
- **Error messaging**: In the event of errors, the sub system needs to inform all involved systems. This can be another subsystem or the machine tool controller itself. For this purpose, programmable logic controller (PLC) variables of the machine tool are used. Each error type is assigned to another PLC variable.

12.2.4 Modelling and Evaluation

After the execution of all machining trials, the autonomous system determines the empirical model coefficients for an arbitrary number of predefined model functions. For this purpose, a generic optimisation algorithm is integrated. Based on the coefficient of determination R^2 as target function, the generic algorithm evaluates iteratively various constellations of model coefficients until the desired model accuracy is achieved. According to Auerbach et al. (2011) the coefficient of determination is a suitable error measure to compare different models with each other. After the determination of the optimised coefficients by a genetic algorithm, the best model is selected by the autonomous system. This is presented to the operator via the visualisation interface.

For the identification of possible model functions, a black-box modelling approach with a symbolic regression has been applied. Symbolic regression allows the approximation of a given data set with the help of mathematical expressions. Thus, it is possible to identify surrogate functions which represent the cause-effect relationships of the investigated machining process. The suitability of the model function with regard to the technological correctness and its complexity has to be evaluated by the technology expert.

12.3 Self-optimised Injection Moulding

In injection moulding the transfer characteristics of the conventional machine control to the process variables can vary by external influences and changed boundary conditions (Fig. 12.5). The conventional injection moulding machine control bases on machine variables. Thus, identical courses of machine variables lead to different process variables in different production cycles. These additional disturbances result in a fluctuating part quality. To increase the process reproducibility the concept of self-optimising injection moulding should compensate occurring process variations.

Fluctuating ambient temperature or varying material properties are systematic disturbances and can affect the product quality heavily. This includes the changes in the heat balance of the injection mould. Fluctuations in the heat balance of the mould occur for example by a non-identical repetitive process such as after changing machine parameters. Therefore, an autonomous parameter adaption has to compensate fluctuations, i.e. in the heat balance of the mould. In contrast to the machine variables process variables provide detailed information about the processes during the injection and holding pressure phase. The cavity pressure path over time for

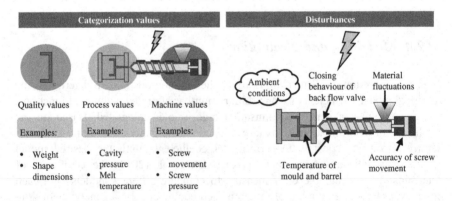

Fig. 12.5 Variables in injection moulding and typical disturbances

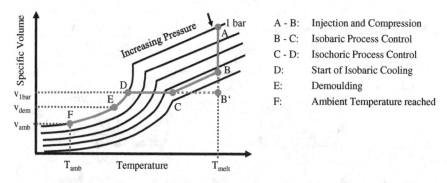

Fig. 12.6 pvT-Diagram as suitable model for the optimisation of holding-pressure phase

example correlates with various quality variables such as the part weight, the part precision, the warpage and the shrinkage, the morphology and sink marks.

Due to the presence of disturbances acting on the injection moulding process, an exclusive control of machine variables does not guarantee an ideal reproducibility of the process and thus constant part properties. Using the pvT-behaviour as a model to map process variables to quality variables, the course of cavity pressure can be adjusted to the actual path of melt temperature. Based on this context, the concept for the self-optimising injection moulding process is derived.

The pvT-behaviour represents the material based interactions between pressure and temperature in the mould of a plastic. It depicts the relationship between pressure, temperature and specific volume and thus allows a description of the link between the curves of cavity pressure, melt temperature and the resulting part properties in injection moulding.

The aim of the self-optimising injection moulding process is to ensure a constant quality of the moulded parts by realising an identical process course in the pvT-diagram (Fig. 12.6). The first requirement is to always achieve an identical, specified specific volume when reaching the 1-bar line (D) in every production cycle. This ensures a constant shrinkage in every cycle. Based on this requirement, the second requirement is to achieve an isochoric process course (C–D), which is characterised by the constant realisation of the given specific volume during the entire pressure phase. Due to the limits of machine the isobaric process control (B–C) is preceded the isochoric process control. Before, the injection and compression phase (A–B) is conventionally controlled by machine values.

At the Institute of Plastics Processing (IKV) a concept for a self-optimising injection moulding machine is being developed. The concept of self-optimising at injection moulding is divided in the MO-System using a model, which is based on the material behaviour, and ISA-Systems, which includes the determination of the melt temperature and a cavity pressure controller (Fig. 12.7).

Based on the conventional injection moulding process the cavity pressure is measured by piezoelectric pressure sensors. The melt temperature is approximated based on the melt temperature in the screw and the mould temperature using the

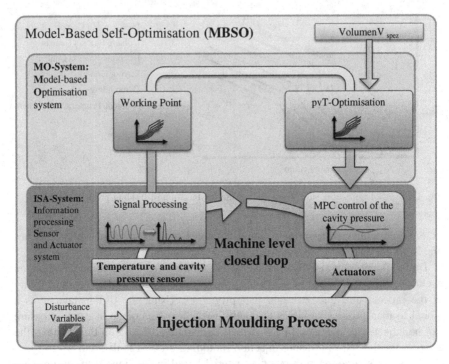

Fig. 12.7 Transfer of the self-optimising concept to the injection moulding process

cooling calculation or directly measured by IR-Sensors (Menges et al. 1980). After determination of the temperature and pressure in the cavity the working point and the optimised trajectory of the pressure can be calculated based on the material specific pvT-behaviour. A Model Predictive Controller (MPC) realises the pressure trajectory autonomously. Using the cavity pressure controller allows to compensate the pressure variations in the cavity. This reduces the natural process variations. Furthermore, the adjustment of the cavity pressure trajectory to the measured temperature in the mould results in the compensation of temperature fluctuations.

To simulate temperature fluctuations the cooling units of the mould are turned off in an experiment after 15 cycles. The temperature path in the mould and the weight of the moulded part is observed using the conventional injection moulding process and the self-optimised concept (Fig. 12.8). Compared to the conventional processing the weight reduction can massively be reduced by using the self-optimised processing concept.

To realise a pvT-optimised injection moulding process the user-friendly implementation of a cavity pressure control is fundamental. The cooperation of the Institute of Plastics Processing (IKV) and the Institute of Automatic Control (IRT) focuses on the autonomous adaption of the cavity pressure control on boundary conditions to simplify the configuration of the cavity pressure controller. Therefore, a dynamic model for a MPC is developed for the injection moulding process (Fig. 12.9).

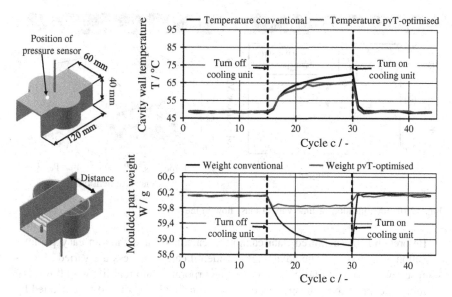

Fig. 12.8 Compensating temperature fluctuations with self-optimising concept compared to conventional processing in injection moulding

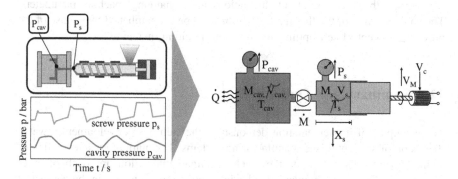

Fig. 12.9 Dynamic model of the MPC to control the cavity pressure

The model describes the correlation of the pressure in the screw (P_s) and the cavity pressure (P_{cav}). Therefore, the system is modelled with two vessels and a valve (Hopmann et al. 2013). To adapt the physical motivated model to the time invariant measurements a time variant parameterisation of the valve is used.

The model is parameterised during an identification cycle. Therefore, a production cycle with a constant screw pressure is realised (Fig. 12.10). Conventionally the screw pressure is controlled in injection moulding. In the current configuration a simple PID-controller is used to realise the constant screw pressure. The difference of the screw pressure to the cavity pressure is measured to detect the mass flow between the vessels over the time. Based on the acquired data a characteristic map is created.

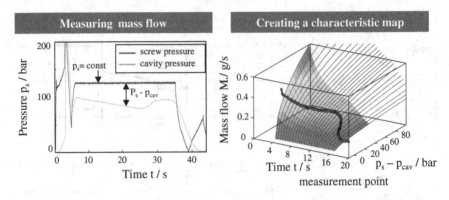

Fig. 12.10 Identification of time variant mass flow in the valve

Beforehand, the acquired data cannot be calculated and thus an easy parameterisation is necessary. The advantages of identification process are varied. A constant screw pressure is feasible and can be incorporated into real-life workflow. The current concept of the self-optimisation injection moulding should be extended by cross-cycle optimisations to counteract disturbances such as viscosity fluctuations. The combination of online control and cross-cycle optimisation is necessary to compensate the heat household fluctuations after changing machine parameters. The compensation of the thermal fluctuations can be accomplished by the use of the previous concept of self-optimising injection moulding machine.

12.4 Summary and Outlook

The examples of implementation demonstrate the step wise development towards the vision of self-optimised manufacturing systems. The autonomous generation of technology models highlights the machine-human interaction approach of automated modelling as the human has a leading and control function in the context of the optimisation system. As by today automated systems are not able to capture all boundary conditions, exceptions and environmental impacts, the machine operator determines the limits and interacts in non-deterministic situations as a decision maker and handles exceptional situations. The automated modelling system enables the development of models by providing an integrated environment for experimental planning by design-of-experiments, deterministic processing of experiments and establishment of machine readable models.

The example of self-optimised injection moulding applies the already known pvT-model for the optimisation of quality features as the specific volume of the moulded parts. The implementation of the model towards a self-optimised production system describes the step-wise procedure to reach optimisation by physically describing the process behaviour and subsequent empirical parameter identification. The result of the optimisation process is a robust process being automatically adapted to temperature fluctuation in the environment. Based on the described work an automated identification process should be possible and further research is conducted on this.

Acknowledgment The authors would like to thank the German Research Foundation DFG for the kind support within the Cluster of Excellence "Integrative Production Technology for High-Wage Countries.

References

Adelt P, Donoth J, Gausemeier J, Geisler J, Henkler S, Kahl S, Klöpper B, Krupp A, Münch E, Oberthür S, Paiz C, Porrmann M, Radkowski R, Romaus C, Schmidt A, Schulz B, Vöcking H, Witkowski U, Witting K, Znamenshchykov (2009) Selbstoptimierende Systeme des Maschinenbaus—Definitionen, Anwendungen, Konzepte. In: Gausemeier J, Rammig FJ, Schäfer W (Hrsg), HNI-Verlagsreihe, Band 234. Heinz Nixdorf Institut, Paderborn
Klaffert T (2007) Selbstoptimierende HSC-Motorspindel. Dissertation, Technische Universität Chemnitz, Verlag wissenschaftliche Scripten, Zwickau
Kahl S (2013) Rahmenwerk für einen selbstoptimierenden Entwicklungsprozess fortschrittlicher mechatronischer Systeme. Universität Paderborn, HNI-Verlagsschriftenreihe, Bd. 308, Paderborn
Klocke F (2014) Technologiewissen für die digitale Produktion. In: Industrie 4.0: Aachener Perspektiven, Tagungsband zum Aachener Werkzeugmaschinenkolloquium 2014, Shaker Verlag, Aachen, p 247–269
Thombansen U, Auerbach T, Schüttler J, Beckers M, Buchholz G, Eppelt U, Gloy Y-S, Fritz P, Kratz S, Lose J, Molitor T, Reßmann A, Schenuit H, Willms K, Gries T, Michaeli W, Petring D, Poprawe R, Reisgen U, Schmitt R, Schulz W, Veselovac D, Klocke F (2012) The Road to Self-optimising Production Technologies. In: Brecher C (ed.) Integrative Production Technology for High-Wage Countries, Springer-Verlag, Berlin Heidelberg, p 793–848
Hopmann C, Reßmann A, Zöller D, Reiter M, Abel D (2013) Strategy for Robust System Identification for Model Predictive Control of Cavity Pressure in an Injection Moulding Process. In: Proceedings of the International Symposium on Measurement Technology and Intelligent Instruments (ISMTII 2013), Aachen, 2013
Menges G, Schmidt L, Kemper W, Storzer A (1980) Rechnerische Beschreibung von Abkühlvorgängen. Plastverarbeiter J 31 (3): 133–136
Klocke F, Buchholz S, Gerhardt K, Roderburg A (2012) Methodology for the Development of Integrated Manufacturing Technologies. In Brecher C (ed.) Integrative Production Technology for High-Wage Countries, Springer-Verlag, Berlin Heidelberg, p 455
Dyckhoff H (1990) A typology of cutting and packing problems. In: European Journal of Operational Research 44: 145–159
Auerbach T, Beckers M, Buchholz G, Eppelt U, Gloy Y-S, Fritz P, Al Khawli T, Kratz S, Lose J, Molitor T, Reßmann A, Thombansen U, Veselovac D, Willms K, Gries T, Michaeli W, Hopmann C, Reisgen U, Schmitt R, Klocke F (2011) Meta-modelling for manufacturing processes. In: Jeschke S, Liu H, Schilberg D (Eds.) ICIRA 2011, Part II, LNAI 7102, Springer-Verlag, Berlin-Heidelberg, p 199–209

Chapter 13
Adaptive Workplace Design Based on Biomechanical Stress Curves

Stefan Graichen, Thorsten Stein and Barbara Deml

Abstract The use of biomechanical models within the fields of workplace and working method design facilitates a detailed consideration of individual physiological capabilities and limitations. Based on motion capturing data of selected manual assembly tasks and the use of a biomechanical body model, body part-oriented stress curves for the upper extremities have been derived. This functional description of physiological stress allows a body part-oriented evaluation of movements and handling positions in the right grasp area. Furthermore these relations have been transferred into body part, movement direction and handled weight dependent linear regression functions. Thereby working system could be enabled to perform physiological stress-oriented self-optimization processes. Applied to manual assembly tasks and in accordance with the individual skills of employees these functions could be the basis for a physiological stress-related adaptive assistant system. Automation engineering, hence, can provide employee-specific support, e.g., in the supply of components or advices for adaption of working method. Working systems thus are able to optimize and adapt themselves to the individual needs and abilities of the employees.

13.1 Introduction

The competitiveness of production systems in high-wage countries can be maintained, among other things, by the development of the self-optimization capability of these systems (Brecher 2011, p. 2). The performance of these socio-technical

S. Graichen (✉) · B. Deml
Institute for Human and Industrial Engineering, Karlsruhe Institute of Technology,
Kaiserstraße 12, 76128 Karlsruhe, Germany
e-mail: stefan.graichen@kit.edu

T. Stein
Institute of Sports and Sports Science, Karlsruhe Institute of Technology,
Engler-Bunte-Ring 15, 76131 Karlsruhe, Germany
e-mail: thorsten.stein@kit.edu

© The Author(s) 2015   175
C. Brecher (ed.), *Advances in Production Technology*,
Lecture Notes in Production Engineering, DOI 10.1007/978-3-319-12304-2_13

systems has to be expanded by adaptive target systems and by reactive systems behaviour (Brecher 2014, p. 3). This requires each element of the production system, i.e. human, technology and organization, to have adaptive capabilities. According to Frank (2004), this means that each element repetitively performs these steps: 1. Analysis of the present situation, 2. Determination of systems objectives and 3. Adaptation of systems behaviour.

Humans are standing in the main focus of this approach. Consequently, the interaction of humans, production technology and organization elements is a topic under comprehensive study. However, these studies focus mainly on cognitive processes (Brecher 2014, p. 65; Mayer 2012; Mayer et al. 2013). Among other things, in the interaction of humans and technology the conformity of technology behaviour with the human operator's expectations is in the focus of these studies (Mayer 2012; Mayer et al. 2013). On top of this, holistic inclusion of humans for the purposes of ergonomic and industrial engineering research also requires direct consideration of the physiological processes involved. This is necessary in order to reach the optimum degree of integration of humans into a production system as demanded by Brecher (2011, p. 796) and, in this way, achieve the required increases in productivity. One possible approach to the inclusion of physiological processes in production systems in real time is presented and its results are discussed below.

13.2 Capabilities of Existing Methods of Workplace Design in Context of Self-optimizing Production Systems

Integrating physiological processes in the self-optimization cycle of a production system requires real time detailed assessment of the physiological status of persons. Several comprehensive approaches are available to analyse and evaluate work processes from a physiological point of view (cf. Hoehne-Hückstädt 2007). In terms of prospective workplace design these procedures allow working systems to be designed in line with the requirements and capabilities of persons (Schaub et al. 2013; Caffier et al. 1999). There are also some approaches explicitly taking into account the individual prerequisites of physiological performance (cf. Sinn-Behrendt et al. 2004). With respect to a discrete working process, procedures are employed either in advance or ex post facto. They are used preventively as well as for correction in the design of working systems. It is not possible to employ these procedures within the framework of production systems with real time capability.

For this purpose, only physiological measurement techniques have been appropriate so far, such as electrocardiograms, electromyograms, or combined methods, such as the CUELA system (Ellegast et al. 2009). The use of these measurement techniques allows data discrete in terms of time to be collected about the physiological strain situation of persons. However, their use needs extensive technical measuring systems, which is possible only to a limited extent in industrial everyday practice and thus restricts the applicability of these procedures.

Consequently, new approaches must be developed which require as little equipment as possible and provide real time indicators of the individual strain status. At this point, above all digital human models are of particular interest. The progressive development specifically of biomechanical human models opens up new possibilities to ergonomic research. This could be a new basis of a more comprehensive inclusion of persons in the process of self-optimization of a production system.

13.3 Use of Biomechanical Human Models for Workplace Design

At the present time, anthropometric and biomechanical human models represent the standard tools of modern workplace design (Bubb and Fritzsche 2009). It is, above all, the biomechanical human models which allow new perspectives to be developed in workplace design. Whether and to what extent these models are able contribute to solutions of open problems in ergonomic research, such as the real time indication of physiological strain, have to be investigated.

In context of this research question, biomechanical human models allow advanced analysis of workplace design and of the interaction of humans and technology more than would be possible with anthropometric models. Models, such as alaska/Dynamicus or the AnyBody Modeling System (Damsgaard et al. 2006), permit in-depth study of the reaction of the musculoskeletal system of persons under the impact of mechanical loads. Workplace design can thus be examined at the level of effects of physiological stresses, and will thereby supply advanced strain indicators (cf. Fritzsche 2010). The use of these models makes it possible to describe, in formal terms, the effects of stresses acting within humans. Compared to the Dynamicus human model, the AnyBody Modeling System (Version 6.02) contains a detailed model of the human muscle system (Damsgaard et al. 2006). It can be used to derive information about the respective physiological strain situation corresponding to the muscle activation computed by the model. Furthermore based on muscle force output an advanced strain indicator is given (Cutlip et al. 2014; Rasmussen et al. 2012). This makes the model suitable for determining effects of physiological stress relative to specific parts of the body.

However, reference must be made at this point also to the respective validity of the model with regard to the muscle forces calculated (cf. Graichen and Deml 2014; Günzkofer et al. 2013; Nikooyan et al. 2010). To bear this fact in mind, the results explained below are interpreted not as relative values, but as values on an ordinal scale (Rasmussen et al. 2012), and only groups of muscles, no individual muscles, are considered.

Nevertheless, there is the possibility to contribute to closing the gap mentioned above in the inclusion of physiological processes within the frame-work of work-place design, and fill this gap in connection with self-optimizing production systems. It has been shown in a current research project that the use of the AnyBody biomechanical human model allows an evaluation of muscle load cycles for specific parts of the body to be performed within the framework of manual assembly processes. Moreover, it was possible to derive biomechanical characteristic curves to determine muscle activation in specific parts of the body. These characteristic curves can be the starting point in assessing a stress situation in real time within the framework of self-optimizing production systems. In combination with a suitable scene recognition, e.g. by means of a Kinect camera system, this makes it possible, in a working process, to determine in real time the change in muscle stresses with respect to specific parts of the body. The working system can thus respond adaptively by indicating individual limits of maximum permissible loads on specific parts of the body. Based on this knowledge the production system could deliver in real time precise advices and specifications to adapt the workplace design or working method.

13.4 Approach for Body Part-Oriented Indication of Physiological Strain in Real Time

In the study, the muscle forces relative to specific parts of the body were calculated on the basis of the AnyBody biomechanical human model. The model represents the human musculoskeletal system as a rigid multi-body model. Using inverse dynamics the model, taking into account interaction forces with the environment and a present kinematics of the individual rigid bodies, calculates the required muscle forces for the considered movements. These constitute the basis from which to derive the biomechanical characteristic stress curves. In line with the anatomical positions of the individual muscles, the muscle groups related to specific parts of the body are set up as follows: forearm (FA), upper arm (UA), shoulder (S), neck (N), and back (B).

The kinematics was logged by an infrared tracking system made by the VICON (MX 13) company with a scanning frequency of 200 Hz. The test setup employed a total of 13 infrared cameras for recording movements. The ground reaction forces were recorded by two AMTI (Advanced Mechanical Technology, Inc., Watertown, MA, USA; 1000 Hz) force plates. For further use in the AnyBody Modeling System (Version 6.02), these data were 2nd-order Butterworth filtered with a cutoff frequency of 12 Hz. Subsequent processing of the data was carried out with the Vicon Nexus 2 (Version 4.6) software.

The test persons were selected on the basis of DIN 33402-2. The selection of test persons followed the anthropometric data of German males 18–25 years old of the 50th percentile with respect to the body height as indicated in this standard. The 16

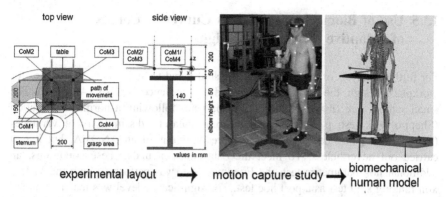

Fig. 13.1 Test setup and modeling in the biomechanical human model

male test persons of the study were all right-handed, had an average body size of 1.784 m (SD 0.013 m) and an average age of 26.6 years (SD 3.2 years).

The study dealt with simple manual assembly movements. In accordance with the MTM basic system (Bokranz and Landau 2011, p. 424) single-handed movement with handling of two different loads (M1 = 1 kg and M2 = 6 kg) have been analysed. The test design included linear movements (length of movement 20 cm) at four points (CoM—Coordinate of Movement) in the right grasp area in three directions positioned orthogonal relative to each other (DoM—Direction of Movement). The experimental layout, a picture from the motion capturing study and the biomechanical model of the AnyBody Modeling System with the applied ground reaction forces and force vector of the handled weight are shown in Fig. 13.1.

As a result of the study it was demonstrated that for the analysed movements activations of muscles relative to specific parts of the body as a function of the length of movement can be described in functional terms by linear regressions ($R^2 > 0.7$). Differentiated by parts of the body (FA, UA, S, N, B), coordinates of movement (CoM 1–4), directions of movement (DoM x, y, z) and weights (M1, M2) to be handled, this was converted into 120 linear regression functions for functional description of body part-oriented stress, in terms of muscle activation, for the execution of the linear movements considered in the right grasp area.

These characteristic curves can constitute an advanced basis of indicating, for specific parts of the body, physiological strain in real time. As a function of the respective position in which an activity is executed, activation of muscles in a specific part of the body can be determined, and the strain acting on these parts of the body can be indicated. Knowing the change in muscle activation as a function of movement and direction can be used to forecast its changes in real time. Either changes in the working method can be derived in order to adjust or reduce a current stress situation and individual strain level, or suitable support functions can be proposed, such as a change in the placement of material at the workplace. The characteristic stress curves constitute the basis of a comparative assessment of strain level encountered at a manual assembly workplace without the need for extensive measuring gear.

13.5 Use of Biomechanical Stress Curves in Context of Adaptive Workplace Design

For further analysis of the findings of the study, and to derive new approaches for workplace design, the data were examined in a variance analysis (ANOVA). The small sample size (n = 16), the data in part not following a normal distribution (Shapiro-Wilk test: $p < 0.05$) and, in some datasets, dissimilarity of variances (Levene test: $p < 0.05$) do violate the preconditions of an ANOVA, but it was carried out nevertheless. To meet the data situation under these conditions, an additional non-parametric test, the Friedman test, was performed with the Wilcoxon-ranking sum test as a post hoc test. The significance level was matched on the basis of the Bonferroni correction ($\alpha = 0.016$). Although its preconditions were violated, the findings of ANOVA were confirmed by the non-parametric test.

Table 13.1 lists the results of the analysis of variances. They indicate the partly significantly different body part related muscle activations. The table shows a comparison of pairs of muscle activation relative to body parts between two directions of movement (DoM: x—to the right, y—to the front, z—to the top) as a function of the coordinates of movement (CoM), body parts, and weights (M1, M2). The direction of movement with the comparatively higher muscle activation is indicated in all cases. The fields marked in colors characterize significant differences (power ≥ 0.7) in direction-dependent muscle activation per body part.

It is evident that, even with the simple movements studied, significant body part related differences in muscle activation can occur. The number clearly increases for weight No. 2. Among the movements studied, most of the significantly higher muscle activations were found in the shoulder and the neck. Moreover, movements

Table 13.1 Differences in body part related muscle activation per coordinate of movement (CoM), direction of movement (DoM), and weight (M)

M	Δ in DoM	CoM 1 body part					CoM 2 body part					CoM 3 body part					CoM 4 body part				
		FA	UA	S	N	B	FA	UA	S	N	B	FA	UA	S	N	B	FA	UA	S	N	B
	X-Y	y	y	x	y	x	x	y	x	x	x	x	y	x	x	x	y	x	x	y	y
M1	X-Z	z	x	x	z	x	z	x	x	z	z	z	z	x	x	z	z	x	x	x	x
	Y-Z	z	y	z	y	y	z	y	z	z	z	z	y	y	z	z	z	y	z	y	y
	X-Y	x	x	x	x	y	x	y	x	x	y	x	x	x	x	y	y	x	x	y	y
M2	X-Z	z	x	x	z	x	z	x	x	z	z	z	x	x	z	z	x	x	z	z	z
	Y-Z	z	z	z	z	y	z	y	z	z	y	z	z	z	z	y	z	z	z	z	y

FA – Fore Arm, *UA* – Upper Arm, *S* – Shoulder, *N* – Neck, *B* - Back

especially in the x- and z-directions result in significantly higher muscle activation than movements in the y-direction.

These findings can now become the basis of an adaptive workplace design. The differences in body part related muscle activation can be employed mainly in self-optimizing systems. On the basis of scene recognition, body part related strain indicators can thus be collected in real time on the basis of the body part-oriented biomechanical stress curves, and filed. If the work process is to include movements with repetitive high muscle activation of the same body parts, it could be assumed that strain level will increase in the body parts concerned, with the result that fatigue can occur in combination with possible discomfort in the execution of the move-ment and changes in movement as a result of fatigue. Consequently, the elements of the musculoskeletal system involved in the movement, such as ligaments, muscles and joints, may be damaged. Allowing the system to intervene at this point, e.g. by changing the working method in accordance with an underlying precedence dia-gram of the assembly task, or by adapting the workplace design with a resultant change in the method of working, can reduce fatigue phenomena resulting from singular body part related muscle activation.

Moreover, the results of the study can constitute the basis of more detailed planning of working methods. As the level of body part related muscle activation is known for the movements considered in the study, these basic types of movement as defined in accordance with MTM, such as "Grasp" and "Bring," can be assigned to specific body part related muscle activities. In defining a working method on the basis of the MTM approach, combinations of basic movements with high activation of identical body parts can be identified and avoided. Besides taking into account parameters of time, distance, and weight, planning of working methods for the first time can also consider the direction-dependent influence on the strain situation of individual body parts of the personnel.

In addition, the outcome of the study can be applied also in taking into account individual performance preconditions in workplace design. On the basis of indi-vidual capability profiles which include restrictions in the execution of specific movements or application of forces, the characteristic load curves allow the working system to be adapted specifically to workers. In this way, an adaptive workplace and method design guided by individual performance preconditions and with real time capability is created which can adapt itself to any situation within the framework of self-optimization of the production system as in a control loop.

13.6 Conclusion and Outlook

The data about body part related indications of strain elaborated in this study can only be a first step in the further integration of the biomechanical human model for workplace design. The biomechanical characteristics of body part related muscle forces initially apply only to the grasp area studied and the group of test persons considered. Consequently, the approach presented here must be extrapolated to other areas of the grasp area and to a heterogeneous group of test persons if a complete description of biomechanical stress functions is to be achieved. This will be the basis of a more far reaching use of the characteristic body part-oriented stress curves within the framework of self-optimizing production systems for a variety of manual work processes.

References

Bokranz R, Landau K (2011) Handbuch Industrial Engineering. Produktivitätsmanagement mit MTM. Band 1, 2., überarbeitete Auflage. Schäffer-Poeschel, Stuttgart

Brecher C (2011) Integrative Produktionstechnik für Hochlohnländer. VDI-Buch. Springer-Verlag, Berlin, Heidelberg

Brecher C (ed) (2014) Exzellenzcluster "Integrative Produktionstechnik für Hochlohnländer". Perspektiven interdisziplinärer Spitzenforschung, 1. Aufl. Apprimus Verlag, Aachen

Bubb H, Fritzsche F (2009) A Scientific Perspective of Digital Human Models: Past, Present, and Future. In: Duffy VG (ed) Handbook of digital human modeling. Research for applied ergonomics and human factors engineering. CRC Press, Boca Raton

Caffier G, Liebers F, Steinberg U (1999) Praxisorientiertes Methodeninventar zur Belastungs- und Beanspruchungsbeurteilung im Zusammenhang mit arbeitsbedingten Muskel-Skelett-Erkrankungen. Schriftenreihe der Bundesanstalt für Arbeitsschutz und Arbeitsmedizin Forschung, vol 850. Wirtschaftsverl. NW, Verl. für Neue Wiss., Bremerhaven

Cutlip K, Nimbarte AD, Ning X, Jaridi M (2014) A Biomechanical Strain Index to Evaluate Shoulder Stress. In: Guan Y, Liao H (eds) Proceedings of the 2014 Industrial and Systems Engineering Research Conference

Damsgaard M, Rasmussen J, Christensen ST, Surma E, de Zee M (2006) Analysis of musculoskeletal systems in the AnyBody Modeling System. Simulation Modelling Practice and Theory 14(8):1100–1111. doi: 10.1016/j.simpat.2006.09.001

Ellegast R, Hermanns I, Schiefer C (2009) Workload Assessment in Field Using the Ambulatory CUELA System. In: Duffy V (ed) Digital Human Modeling, vol 5620. Springer Berlin Heidelberg, pp 221–226

Frank U, Giese H, Klein F, Oberschelp O, Schulz B, Vöcking H, Witting K (2004) Selbstoptimierende Systeme des Maschinenbaus. Definition und Konzepte, Paderborn

Fritzsche F (2010) Kraft- und haltungsabhängiger Diskomfort unter Bewegung—berechnet mit Hilfe eines digitalen Menschmodells. Dissertation, Technische Universität München

Graichen S, Deml B (2014) Ein Beitrag zur Validierung biomechanischer Menschmodelle. In: Jäger M (ed) Gestaltung der Arbeitswelt der Zukunft. 60. Kongress der Gesellschaft für Arbeitswissenschaft; TU und Hochschule München 12.–14. März 2014. GfA-Press, Dortmund, pp 369–371

Günzkofer F, Bubb H, Bengler K (2013) The validity of maximum force predictions based on single-joint torque measurements. In: 2nd International Digital Human Modeling Symposium

Hoehne-Hückstädt U (2007) Muskel-Skelett-Erkrankungen der oberen Extremität und berufliche Tätigkeit. Entwicklung eines Systems zur Erfassung und arbeitswissenschaftlichen Bewertung von komplexen Bewegungen der oberen Extremität bei beruflichen Tätigkeiten, 1. Aufl. BGIA-Report, vol 2007, 2. Technische Informationsbibliothek u. Universitätsbibliothek; HVBG, Hannover, Sankt Augustin

Mayer MP (2012) Entwicklung eines kognitionsergonomischen Konzeptes und eines Simulationssystems für die robotergestützte Montage. Industrial engineering and ergonomics, Bd. 12. Shaker, Aachen

Mayer MP, Schlick CM, Müller S, Freudenberg R, Schmitt R (2013) Systemmodell für Selbstoptimierende Produktionssysteme. Kognitive Systeme(1)

Nikooyan AA, Veeger, H. E. J., Westerhoff P, Graichen F, Bergmann G, Van Der Helm, F. C. T. (2010) Validation of the Delft Shoulder and Elbow Model using in-vivo glenohumeral joint contact forces. Journal of Biomechanics 43(15):3007–3014. doi: 10.1016/j.jbiomech.2010.06.015

Rasmussen J, Boocock M, Paul G (2012) Advanced musculoskeletal simulation as an ergonomic design method. Work: A Journal of Prevention, Assessment and Rehabilitation 41 (0):6107–6111. doi: 10.3233/WOR-2012-1069-6107

Schaub K, Caragnano G, Britzke B, Bruder R (2013) The European Assembly Worksheet. Theoretical Issues in Ergonomics Science 14(6):616–639. doi: 10.1080/1463922X.2012.678283

Sinn-Behrendt A, Schaub K, Landau K (2004) Ergonomisches Frühwarnsystem "Ergo-FWS". In: Landau K (ed) Montageprozesse gestalten. Fallbeispiele aus Ergonomie und Organisation. Ergonomia-Verl., Stuttgart, pp 233–248

Part VI
Human Factors in Production Technology

Part VI
Human Factors in Production Technology

Chapter 14
Human Factors in Production Systems

Motives, Methods and Beyond

Philipp Brauner and Martina Ziefle

Abstract Information and communication technology (ICT) is getting smaller and faster at a dashing pace and is increasingly pervading production technology. This penetration of ICT within and across production technology enables companies to aggregate and utilize massive amounts of data of production processes, both horizontally (across different products) and vertically (from machine level, over the shop floor, to the supply chain level). Presumably, this yields in Smart Factories with adaptable manufacturing processes that adjust to different goals, such as performance, product quality, or resource efficiency. But the increasing amount of available data also raises considerable challenges: Strategic decisions still depend on humans in the loop who have to perceive and process increasingly complex multi dimensional data sets and to make decisions whose effects are increasingly difficult to forecast. This paper is concerned with the potential of the human factor. Along with three case studies, we demonstrate the potential of human factors in the development of applications for Smart Factories and enterprises in the era of Industry 4.0. The paper concludes with a set of guidelines and methods for the user-centred development of applications for Industry 4.0.

Keywords Human factors · Socio-Technical systems · User diversity · Industry 4.0

P. Brauner (✉) · M. Ziefle
Human-Computer Interaction Center, RWTH Aachen University,
Campus-Boulevard 57, 52056 Aachen, Germany
e-mail: brauner@comm.rwth-aachen.de

M. Ziefle
e-mail: ziefle@comm.rwth-aachen.de

© The Author(s) 2015
C. Brecher (ed.), *Advances in Production Technology*,
Lecture Notes in Production Engineering, DOI 10.1007/978-3-319-12304-2_14

14.1 Motives for Integrating Human Factors in Production Engineering—the Challenge

Production systems are not like they used to be. The 21st century will confront enterprises and manufacturing companies with completely novel generations of technologies, services, and products based on computer technologies (Schuh and Gottschalk 2008; Schuh et al. 2009). In order to meet competition on global markets and to ensure long-term success, the companies need to adapt to shorter delivery times, increasing product variability and high market volatility, by which enterprises are able to sensitively and timely react to continuous and unexpected changes (Wiendahl et al. 2007). One of the major cornerstones to meet these challenges is the implementation of digital information and communication technologies into production systems, processes and technologies, which allow novel developments by combining the physical world and fast data access and data processing via the Internet (Industry 4.0).

Another major cornerstone is to understand the impact of the human factor and to integrate human factors knowledge seamlessly in the technology development cycle, thus moving from traditionally purely technical systems into socio-technical systems.

In the next decades new generations of technology systems and products have to master fundamental societal and technological challenges (Wilkowska and Ziefle 2011). This includes the impact of the greying society, with an increasingly aged work force, but also short technological life cycles triggered by fast changing technological systems and the question in how far diversely skilled workers might learn and adapt to the increasing complexity of systems (Ziefle and Jakobs 2010). Although the crucial potential of usable products that are appropriate for a diverse user group, recognition of the importance of diversity is only slowly influencing mainstream technology development practise. New approaches integrate users as a valuable source for new ideas and innovations (end-user driven innovation cycle) and integrate their knowledge as an integral component into the technical development (Franke and Piller 2004). User communities are a significant source for innovation and provide market insight before launching an innovative product (Fredberg and Piller 2011).

For high-wage countries, which are characterized by competitive production systems and a high pressure to succeed, it is more than high time to integrate human factors knowledge as a natural and expert source of information into the technology development and processing.

We conclude that these challenges can only be addressed if methods from production engineering are "reinvented" and combined with methods from the social sciences. Only a holistic inter- and transdisciplinary methodology will be able to address the changing production processes and changing workforce, in order to strengthen the competitiveness of companies in high-wage countries (Calero Valdez et al. 2012).

14.1.1 The Contribution of the Social Sciences

Traditionally, social science research deals with the understanding of the human factor, i.e. explaining, measuring and predicting human experience, affective states, cognitions, and behaviour. This includes the capturing of emotional states and cognitive abilities in different application contexts, against the background of user diversity and the differences in attitudes within and across individuals. But it covers also developmental changes of humans across the life span as well as the systematic understanding, control and change of human behaviours.

The huge volatility of the human nature and its enormous adaptability to different situations necessitated the development and establishment of standardized human factors' methods and metrics that allow a sound, reliable and valid prediction of the human factor. Hence, on the base of their profound empirical methodology and their sound knowledge base of the human factor, social science can naturally contribute to the understanding of socio-technical systems, as e.g. the current challenges in production technology.

Basically, social science knowledge can contribute to different challenges:

- *Dealing with Complexity* Understanding human cognition and decision making in production systems
- *Meeting Requirements of User Diversity* Understanding the impact of the demographic change and diversely skilled workers
- *Handling of Heterogeneous Teams* Understanding the interaction and communication of impacts by interdisciplinary working environments and team culture
- *Measuring Technology Acceptance* Understanding the benefits and caveats of the interaction and communication of humans with technology (from (hybrid) systems to interfaces)
- *Dealing with Usability and User Experience* Understanding the impact of respecting human cognitions, emotions, expectations and values in the development of technology
- *Providing Experimental and Empirical Metrics* Understanding novel contexts and potential settings (from prototypes to real settings) by applying testing scenarios that allow evaluating the quality of measures and settings.

With regard to work environment and design of production systems, usability and user experience research has been shown to exploit a huge benefit for diverse application setting within industrial environments. In many domains (perceived) usability is an established criterion for the quality of a product and is increasingly a deal breaker for the buying intention for consumer electronics. However, usability is not yet widely acknowledged for professional applications in production engineering (Myers et al. 1996; Ziefle and Jakobs 2010) and many interfaces suffer from complicated screen layouts (Arning and Ziefle 2009; Ziefle 2010a), not understandable labels and buttons, undistinguishable and unclear icons (Pappachan and Ziefle 2008), and steep learning curves. Interfaces that somehow work for specific user groups, but fail for the increasingly changing and increasingly diverse labour force (Wilkowska and Ziefle 2011; Calero Valdez et al. 2013).

The reason for investing in usability is simple: Good soft- and hardware usability is the reason for a higher quality of the achievements and increased productivity of the workers (Arning and Ziefle 2007; Ziefle 2010b). Also, good interfaces reduce the time and costs for training and support and increases the users' work satisfaction, loyalty and commitment (Arning and Ziefle 2010). Furthermore, if usability is considered from early on in the design process and not an added on top in the final stages of the development, it can dramatically decrease the development time and costs (Nielsen 1993; Holzinger 2005). Hence, establishing usability as a core criterion for professional applications can increase the return of investment (ROI) and decrease the total costs of ownership (TCO) dramatically.

14.2 Methods for Understanding and Quantifying Human Factors—the Potential

The following section presents a bouquet of measures that can be applied in various stages of the development process and how efficient, effective and usable systems can be realized for a changing and increasingly diverse user population.

ISO/EN 4291/11 defines *effectiveness, efficiency, and user-satisfaction* as the three central criterions for the usability of interactive products. Although this norm provides essential usability criteria, it neither presents methods for realizing systems with high usability, nor does it provide off-the-shelf measures to quantify the usability of systems. So how can user-friendly systems be realized?

The foundations for designing usable systems for people can be found in the early dawn of graphical user interfaces: In 1985—just a year after Apple introduced Macintosh—Gould and Lewis proposed three key principles for building usable software (Gould and Lewis 1985): *Early focus on the users tasks, empirical measurement of the product usage* and *iterative design*. Many stakeholders are involved in the specification and development of new software systems: Software engineers, user interface designers, and software architects on one side, as well as domain experts and managers who are responsible for the introduction of the new software. Of course, the end users also belong in the circle of stakeholders, but often enough they are neglected or consulted only in the latest stages of the development.

Early focus on the users tasks The decision to invest in new software often comes from the managers of a company or external consultants. As they are not the actual users this often leads to miss defined or insufficiently defined task descriptions. Hence, users must be included in the earliest stages of the design of a system, their tasks must be well understood and the users' wishes, abilities and motives must be captured and well understood by the design team.

Empirical measurement of the product usage To ensure that the developed system supports the tasks users want or have to perform the execution of prototypic

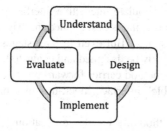

Fig. 14.1 Schematic presentation of an iterative development process as proposed by Gould and Lewis (1985)

tasks by actual users must be observed. These task executions can be quantified (e.g. by measuring task completion, accuracy) and will identify the critical parts of the software that do not support users in their work and need to be revised.

Iterative design It is not sufficient to attest a certain level of usability at certain point in time, but system usability must be a continuous focus during the whole design process. Ideally, a design team starts with an early prototype of a planned system and evaluates the usability of the system with typical tasks performed by typical users. These quantitative evaluations eliminate the most severe usability issues early in the development. Then, future iterations of the system with functional prototypes can focus on other and minor usability issues, but again with actual users who perform typical tasks with the planned system. Figure 14.1 shows a schematic presentation of this cyclic process.

14.2.1 *Metrics, Procedures and Empirical Approaches*

This section outlines a few but still the most central usability methods for user-centred design. An in-depth description and additional methods can be found in Dix et al. (2003) or Courage and Baxter (2005).

Methods for developing and evaluating user interfaces can be divided in methods with and methods without involvement of prospected users. Methods without user involvement, such as *heuristic evaluation* (when experts evaluate if interfaces meet certain heuristics), *GOMS* (a method for estimating the expected performance, similar to Methods-Time Measurement—MTM), or *cognitive walk-throughs* (human factors experts identify usability issues by predicting how users would solve tasks) are not covered in this article. They are a valuable addition to every development process, yet many issues will remain uncovered, if prospected users are not included during the design phases.

Paper prototyping Paper prototyping is a low fidelity prototyping technique that allows the gathering of user feedback on an interface in early stages of the design and before the software implementation starts. The proposed user interface is drawn on paper and discussed with users. Their feedback can be integrated at once and interface suggestions can be drawn and discussed immediately. Replacing parts of the simulated screen with new layers can simulate interactive interfaces and the traversal through multiple screens. Paper prototyping is best applied in early stages of the design, when the general interface is designed. But even in later stages individual changes, new dialogs and screens can quickly be designed and evaluated using this technique.

Rapid prototyping This method carries forward the concept of paper prototyping. This may include "clickable" interface mock-ups in a presentation tool, or functional screen prototypes.

Wizard-of-Oz This method allows the evaluation of interfaces even if the respective backend functionality is not yet available, by observing the user's interaction with the prototyping system and simulating its outcome. For example, a paper prototype can "control" a machine, if a hidden observer simulates the user's interaction with a remote control.

AB-tests To understand if different interface alternatives result in higher speed, higher accuracy or higher user satisfaction different design alternatives can be compared by presenting both to a set of users, either within-subjects (every user uses every interface) or between-subjects (each user evaluates just one interface). Measures may be objective measures, such as task performance and learnability, or subjective measures, such as users' perceived efforts.

Note that the feedback acquired changes with the perceived completeness of the interface: Paper prototypes are perceived as easy to change, thus users often request fundamental changes, while applications that appear complete are apprehended as difficult to change, hence the articulated feedback is often restricted to wording, colour choices, and other trivialities.

The methods presented are a necessity for building usable, understandable, learnable, goal-oriented, task-adaptive, efficient and satisfying software for production systems. They should as a matter of fact belong to the standard toolbox of all stakeholders included in the technical development process. To identify and obliterate usability pitfalls and to assure that the software captures the users actual needs, these and similar methods must be applied frequently during the design process. However, to unfold their full potentials, human factors experts must also be included in this process, as they are able to detangle individual differences in effectiveness, efficiency and user-satisfaction that are caused for example by motivation, personality, or cognitive abilities. These then allows a fine-grained tuned or individually tailored user interfaces.

14.2.2 Case Studies—Examples of the Potential of Exploring Human Factors

The following three exemplary cases present "success stories" in which the methodology of user-centred design and human factors research was applied in the area of production systems.

The first case quantifies the influence of poor usability on efficiency while interpreting large data sets in supply chain management. The second case outlines how human factors relevant for good job performance in supply chain management can be identified. The third case describes how an adequately designed worker support system can relocate the focus between speed and accuracy depending on the task.

Case 1—Visual and Cognitive Ergonomics of Information Presentation

Managing the follow of material in supply chain management depends on both, the ability to perceive, understand and interpret given data correctly, as well as the presentation of the data. To understand how insufficient data presentation and bad usability impacts the decision quality, we conducted a formal *AB-test*. We measured the decision speed and decision quality in dependence on human information processing speed and the presentation form (poor and good usability constructed as small or medium font sizes in a tabular presentation of supply chain data). The study revealed that poor usability obviously decreases the overall performance. Strikingly though is the finding, that poor usability disproportionately impairs people who have a lower information processing speed, while faster information processors can compensate poor information presentation. This finding highlights the necessity of user-centred and participatory design, as software developers, interfaces designers and contributing mechanical engineers usually do not realize the negative effects of poor interfaces on decision speed and decision quality, as they are able to compensate the negative effects, while the end users often are not. Frequent user tests with methods as presented above will reveal these barriers even in early stages of the design.

Case 2—A game-based Business Simulation to Understand Human Factors in Supply Chain Management

Supply chains are sociotechnical systems with high dimensional and nonlinear solution spaces. The performance of a supply chains is not only determined by technical factors (e.g. shipping times, replacement times, delivery strategies, lot sizes, order costs, etc.) but also by the abilities of the human operators who needs oversee the possible choices and make good decisions in this complex solution space. To identify the factors that contribute to a better understanding of the supply chain and to develop methods that help supply chain managers to make better decisions in shorter time we developed a series of supply chain games (Brauner et al. 2013; Stiller et al. 2014). The business simulation games are virtual test beds with multiple uses: First, they are a flexible tool to identify and quantify human factors that contribute to efficiency and effectivity in managing information in logistics and supply chain management, by systematically varying the difficulty of the game and investigating how different human factors (capacity, processing speed, motivation, self-efficacy, personality traits) relate to game performance. Second, through experimental variation the user interface and/or provided decision support tools, the benefit or costs of these can be evaluated and quantified within the test bed before the proposed changes are implemented in commercial applications. Third, the relationship between in-game performance and performance in the job as supply chain mangers can be used to develop interactive methods for personnel selection with an increased accuracy.

The design of this research and evaluation framework followed the design principles presented above and several of the usability methods presented above were applied. The following sections describe the iterative development process and some of the methods used during the process.

The development of the business simulation game was a collaborative effort between the four disciplines mechanical engineering, communication science, computer science and psychology. Each of the four disciplines contributed methods in order to on ensure the simulation model's validity on one side and the good usability and suitability for psychometrical evaluations on the other side.

At the beginning of the project the experts form each discipline discussed the game model and the relevant indicators for inferring the simulated companies status. Then the *paper prototyping* technique was used to arrange the indicators to form the user interface of the game. In a second step a low-fidelity software prototype of the game simulation was realized and the previously selected indicators were populated with data from the simulation model and the experts evaluated the suitability of the indicators and the simulation model. Third, the game was implemented as a web application and the design of the user-interface ware strictly based on the earlier prototypes and influenced by technical considerations. Forth, during one user study feedback from external usability experts was gathered and their suggestions were integrated in the game. A subsequent user study attested that the user interface refinements led to an increased profit of the simulated company as the users hat a better overview of the performance indicators and were able to make better decisions. Throughout the design process feedback was gathered from other experts and test users and the user interface and the game model was refined accordingly. Figure 14.2 shows the development progress of the game across three prototype levels.

Case 3—Augmented Reality Worker Support Systems

The third test case of including human factors research and design methodologies in production engineering is the design and evaluation of a work support system for carbon-fibre reinforced plastic manufacturing (CFRP) (Brauner et al. 2014).

The CFRP production process relies on manual production steps in which multiple layers of carbon fibre cloth have to be aligned in specific orientations. The overall stability of a CFRP part is prone to misalignments and a mismatch of 5° reduces the mechanical stability of a component by 50 %. These defects can only be

Fig. 14.2 Different development stages of the Supply Chain Simulation Game (*left* paper prototype, first layout of the interface, *centre* rapid prototype in a spread sheet applications for evaluating the game's model, *right* final user interface of the game)

detected late in the production process, which yields in extra costs for the process steps between origin of the error and the detection.

To increase the stability of the process we designed a worker support system that alleviates the variances in the orientations of the carbon fibre cloths. The system captures the orientation of the currently placed layer and provides auditory and visual feedback to the worker. Later iterations of the prototype may be realized stationary in the assembly cell or on mobile augmented reality systems, such as Google Glass.

An evaluation of a *rapid prototype* of the system with the *Wizard-of-Oz technique* (compared four different feedback modalities against each other (no feedback, auditory, visual, and combined feedback). The key finding is that providing no feedback yielded in the highest speed and the lowest accuracy, while combined auditory and visual feedback led to slowest speed and highest accuracy. Hence, a target-oriented design of worker support systems can nudge workers to either increase the speed of the production process or to increase its accuracy (speed—accuracy trade-off).

Summarizing all three cases, utilizing methods from user-centred design and human factors research reveal a deeper understanding of the behaviour of human workers on different levels of production processes. The methods facilitate the development of better systems and applications for production processes in shorter time. Adequately designed system can influence the production processes by shifting the trade-off between speed and accuracy by giving carefully designed feedback. Furthermore, some methods allow an explanation and prediction of individual differences in speed, accuracy, performance and motivational factors that might either contribute to better designed and better targeted software systems or to more specialized recruiting and training processes for the employees.

14.3 Beyond—How to Amend Productivity with Quality of (Work)Life—the Vision

The truly understanding and consideration of human factors for the realisation of human and humane working environments is a critical issue. Though it might be still more critical when facing the upcoming generation Y (Martin 2005). According to Bakewell and Mitchell (2003), this generation can be characterized along five types: the *"recreational quality seekers"*, the *"recreational discount seekers"*, the *"trend setting loyals"*, the *"shopping and fashion uninterested"* and the *"confused time/money conserving"*. This new work force generation (born after 1977) brings thus another working and performance attitude to high wage countries. It is less the mere concentration on work or system performance alone, only addressing pragmatic and productivity aspects of technology, it is far more, value-oriented, hedonic and highly fulfilling working conditions that characterize the challenges of high performance cultures in the near future.

Two keystones for this reformation of productivity might be shortly outlined, one refers to a more technical one, the pattern language to enable interdisciplinary teams, the other one is more visionary and relates to the working climate of the future Generation Y.

14.3.1 Enabling Communication in Interdisciplinary Teams

A common problem among interdisciplinary teams is the lack of a shared language and misconceptions about the other methodologies. Established systems for enabling interdisciplinary communication in teams are *pattern languages*. Christopher Alexander's seminal work suggested these languages as a method to enable different stakeholders in urban planning (e.g. architects, civil engineers, city planers and residents) to collaboratively design the living space (Alexander et al. 1977). Each pattern describes a solution for recurring problems, defines a shared iconic name, captures the forces that argue for and against the given solution and refer to other patterns that relate to the solution, either as possible alternatives (horizontal) or superordinate and subordinate patterns (vertical). A network of interlinked patterns then forms a complete pattern language. Other disciplines adopted pattern languages as a tool to capture disciplinary knowledge, but sacrificed the aspiration for participatory design. In computer science the "Gang of Four" introduced software design patterns (Gamma et al. 1995) that quickly revolutionized the communication between experts in software engineering. In mechanical engineering (Feldhusen and Bungert 2007) suggested a pattern language to manage archetypal engineering knowledge, but again, this pattern language captured engineering knowledge by and for experts. It shows that pattern languages exist for various domains, but still Alexander's original goal to enable all stakeholders to jointly develop holistic solutions got astray.

Thus, a dedicated pattern language for the design of production systems may enable truly participatory design in production engineering and a more efficient collaboration among interdisciplinary teams. The goal of this language must be to empower all stakeholders to understand the constraints of a given problem and overview the set of possible solutions. This language could cover individual competencies and methods of the contributing disciplines and will enable interdisciplinary teams to collaborate more efficiently on future production systems.

14.3.2 Motivators for High Performance Cultures

Recurring again to the generation Y, the novel attitude of workers might also requests a change within the performance culture in the production and work environment. In this perspective, the quality of "good interface of technology" relies on affective and hedonic aspects of work and production—attributes emphasizing individuals' well-being, pleasure and fun when interacting with technology and technological systems, the quality and the design of products, but also the well-being of teams, working groups as well as the well-being of society, focusing on social morality, working ethics, work-life balance, environmental justice, or life style. To this end, the relationship of users and technological products and their working environment is of importance and the making sense of user experience. In addition, the work experience and domain knowledge of workers, end users and consumers of technology or technical systems is of high value (and is so far, mostly ignored). It seems indispensable for efficient production environments to focus on human factors in order to enable highly motivated and high performance teams not only steering with the traditional motivators—money, pressure, or competition—but rather to focus on the internal motivation of workers to contribute to the system effectiveness by including their knowledge and their expertise within iterative product development cycles.

Naturally, the relationship between leaders and workers need to be reformatted accordingly. Efficient teams then should be characterized by transparent group communication, a commonly shared information policy and the appreciation of ideas and innovations created by working teams. This team culture though requires trust in both, the team leader and the workers and might be a overdue performance driver of current working environments in enterprises that might ensure sustainable high performance cultures on the long run.

Acknowledgments The authors thank Sebastian Stiller, Robert Schmitt, Malte Rast and Linus Atdorf for lively and successful collaboration. We also owe gratitude to Frederic Speicher, Victor Mittelstädt and Ralf Philipsen who contributed to this work. The German Research Foundation (DFG) funded the presented projects (Cluster of Excellence "Integrative Production Technology for High Wage Countries").

References

Alexander C, Ishikawa S, Silverstein M (1977) A Pattern Language: Towns, Buildings, Construction. Structure 2:1171. doi: 10.2307/1574526.

Arning K, Ziefle M (2009) Different Perspectives on Technology Acceptance : The Role of Technology Type and Age. HCI and Usability for eInclusion 5889:20–41. doi: 10.1007/978-3-642-10308-7_2.

Arning K, Ziefle M (2007) Understanding age differences in PDA acceptance and performance. Computers in Human Behavior 23:2904–2927.

Arning K, Ziefle M (2010) Ask and You Will Receive. International Journal of Mobile Human Computer Interaction 2:21–47. doi: 10.4018/jmhci.2010100602.

Bakewell C, Mitchell V-W (2003) Generation Y female consumer decision-making styles. International Journal of Retail & Distribution Management 31:95–106.

Brauner P, Bremen L, Ziefle M, et al (2014) Evaluation of Different Feedback Conditions on Worker's Performance in an Augmented Reality-based Support System for Carbon Fiber Reinforced Plastic Manufacturing. In: Ahram T, Karwowski W, Marek T (eds) Proceedings of the 15th International Conference on The Human Aspects of Advanced Manufacturing (HAAMAHA): Manufacturing Enterprises in a Digital World. CRC Press, Boca Raton, pp 5087–5097.

Brauner P, Runge S, Groten M, et al (2013) Human Factors in Supply Chain Management— Decision making in complex logistic scenarios. In: Yamamoto S (ed) Proceedings of the 15th HCI International 2013, Part III, LNCS 8018. Springer-Verlag Berlin Heidelberg, Las Vegas, Nevada, USA, pp 423–432.

Calero Valdez A, Schaar AK, Ziefle M, et al (2012) Using mixed node publication network graphs for analyzing success in interdisciplinary teams. Lecture Notes in Computer Science (including subseries Lecture Notes in Artificial Intelligence and Lecture Notes in Bioinformatics). pp 606–617.

Calero Valdez A, Schaar AK, Ziefle M (2013) Personality influences on etiquette requirements for social media in the working context. When jaunty juveniles communicate with serious suits. Human Factors in Computing and Informatics LNCS 7946. Springer, Berlin, pp 431–450.

Courage C, Baxter K (2005) Understanding Your Users: A Practical Guide to User Requirements Methods, Tools, and Techniques. Morgan Kaufmann Publishers.

Dix A, Finlay J, Abowd GD, Beale R (2003) Human Computer Interaction, 3. edn. Pearson.

Feldhusen J, Bungert F (2007) Pattern Languages: An approach to manage archetypal engineering knowledge. ICED07: 16th International Conference of Engineering Design. Paris, France, pp 581–582.

Franke N, Piller F (2004) Value creation by toolkits for user innovation and design: The case of the watch market. Journal of Product Innovation Management 21:401–415. doi: 10.1111/j.0737-6782.2004.00094.x.

Fredberg T, Piller FT (2011) The paradox of tie strength in customer relationships for innovation: A longitudinal case study in the sports industry. R&D Management 41:470–484. doi: 10.1111/j.1467-9310.2011.00659.x.

Gamma E, Helm R, Johnson R, Vlissides J (1995) Design Patterns. Addison-Wesley Professional.

Gould JD, Lewis C (1985) Designing for Usability: Key Principles and what Designers Think. Communications of the ACM 28:300–311. doi: 10.1145/3166.3170.

Holzinger A (2005) Usability engineering methods for software developers. Communications of the ACM 48:71–74. doi: 10.1145/1039539.1039541.

Martin CA (2005) From high maintenance to high productivity: What managers need to know about Generation Y. Industrial and Commercial Training 37:39–44.

Myers B, Hollan J, Cruz I, et al (1996) Strategic Directions in Human-Computer Interaction. ACM Computing Surveys 28:794–809. doi: 10.1145/242223.246855.

Nielsen J (1993) Usability Engineering. Morgan Kaufmann Publishers Inc., San Francisco, CA, USA.

Pappachan P, Ziefle M (2008) Cultural Influences on the Comprehensibility of Icons in Mobile-Computer-Interaction. Behaviour and Information Technology 27:331–337.

Schuh G, Gottschalk S (2008) Production engineering for self-organizing complex systems. Production Engineering 2:431–435.

Schuh G, Lenders M, Nussbaum C, Kupke D (2009) Design for changeability. Changeable and Reconfigurable Manufacturing Systems. Springer, London, pp 251–266.

Stiller S, Falk B, Philipsen R, et al (2014) A Game-based Approach to Understand Human Factors in Supply Chains and Quality Management. Proceedings of the 2nd International Conference on Ramp-Up Management 2014 (ICRM). Elsevier B.V., p (in press).

Wiendahl HP, ElMaraghy HA, Nyhuis P, et al (2007) Changeable Manufacturing—Classification, Design and Operation. CIRP Annals—Manufacturing Technology 56:783–809. doi: 10.1016/j.cirp.2007.10.003.

Wilkowska W, Ziefle M (2011) User diversity as a challenge for the integration of medical technology into future home environments. In: Ziefle M, Röcker C (eds) Human-Centred

Design of eHealth Technologies. Concepts, Methods and Applications. IGI Global, Gersgey, P. A., pp 95–126.

Ziefle M (2010a) Information presentation in small screen devices: The trade-off between visual density and menu foresight. Applied Ergonomics 41:719–730.

Ziefle M (2010b) Modelling mobile devices for the elderly. Advances in Ergonomics Modeling and Usability Evaluation.

Ziefle M, Jakobs E-M (2010) New challenges in human computer interaction: Strategic directions and interdisciplinary trends. 4th International Conference on Competitive Manufacturing Technologies. University of Stellenbosch, South Africa, pp 389–398.

Chapter 15
Human Factors in Product Development and Design

Robert Schmitt, Björn Falk, Sebastian Stiller and Verena Heinrichs

15.1 Introduction

The design and operation of product development processes are typical problems in engineering domains and quality management. In order to guarantee the efficient and effective realization of products various methods and tools have been developed and established in the past. Nevertheless these methods where usually designed to fit engineer-to-cost strategies for cost efficient products. Recent success stories of companies in different industrial sectors have proven, that engineer-to-value strategies can lead to an even higher profitability of products due to higher margins when combining value and cost orientation (ISO/IEC: 15288:2008; Schuh 2012). The two levers for engineer-to-value product management are illustrated in Fig. 15.1.

One of the major key elements for a successful introduction of engineer-to-value product management is the introduction and consideration of human factors. Therefore the integration of the customer in product development processes and the detailed analysis of the customer perception are essential (Brecher et al. 2014). The aspects concerning the role of customers in the product development will be presented in the first and second chapter of this paper. The third chapter will change the focus on the company perspective and will discuss new approaches and possible solutions how to develop products in a more efficient and human oriented way.

R. Schmitt · B. Falk · S. Stiller (✉) · V. Heinrichs
Laboratory for Machine Tools and Production Engineering (WZL) of RWTH
Aachen University, Steinbachstr. 19, 52074 Aachen, Germany
e-mail: s.stiller@wzl.rwth-aachen.de

© The Author(s) 2015
C. Brecher (ed.), *Advances in Production Technology*,
Lecture Notes in Production Engineering, DOI 10.1007/978-3-319-12304-2_15

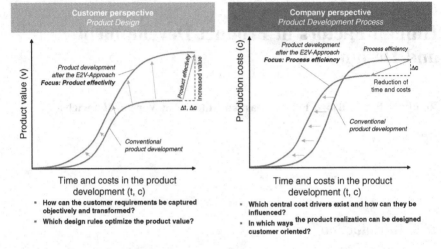

Fig. 15.1 Engineer-to-Value combines the effect of higher prices through an increase of product value while reducing the costs and time for product development and realization

15.2 The Human Perception of Quality

Customer satisfaction is a major component of a company's reputation and economic success. The relationship between a company and its customer base is established by and related to the quality of the products provided by the company. The degree to which customers are satisfied with a product results from the perceived quality of the product in question. The perceived quality of a product is the outcome of a cognitive and emotional mapping-process between the customer's conscious and unconscious experience, his or her expectations towards the product in theory and his or her experience with the product in practice (Schmitt et al. 2009). The amount of customer satisfaction correlates with the extent to which the product exceeds, fulfills or disregards the customer's requirements during real-life application (Homburg 2008). The disregard of customer requirements during product development and design leads to the customer's denial of the product resulting in a decreased willingness to pay. Furthermore, the exceedance of customer requirements (*over engineering*) is not profitable since customers hesitate to invest more money into product features which they do not necessarily require (see Fig. 15.2) (Masing 2007).

The degree of a customer's satisfaction with a product's quality changes during the progress of real-life application. The changes are due to a decrease or increase of the perceived value of the product over time whereupon the time period before and after the purchase designates the relevant time interval. During the pre-purchase phase, the customer gathers information about the product consulting family members and friends while simultaneously browsing through commercial web pages and social media platforms. The post-purchase phase refers to the period

Fig. 15.2 The impact of the degree of fulfillment of customer requirements on the customer's willingness to pay (Value function; reproduced from Masing (2007))

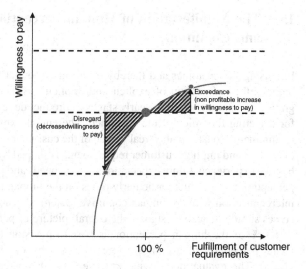

of product usage and application (Schmitt and Amini 2013). The perceived value of a product consists of 13 unequally weighted factors such as brand, design, practicability, purchase price and follow-up costs. The customer perceived value (CVP) quantifies the trade-off between the customer's perceived quality of a product and the customer's perceived invest into the product (see Fig. 15.3) (Schmitt et al. 2014).

The extent to which customer requirements are met in product development and design is critical to a product's market acceptance and success. The integration of customers into the early stages of product development and design becomes more and more important with respect to market leadership and differentiation, especially in flooded markets.

Fig. 15.3 Change of CVP during tablet use and application. The presented values represent the mean values of 36 test persons (own material)

15.3 The Manifestation of Human Perception and Cognition

Enthusing the customer and thereby ensuring a product's level of competitiveness requires the satisfaction of explicit and implicit customer requirements. The integration of customers into the early stages of product development and design allows for detecting the voice of the customer, whereupon explicit and implicit customer requirements constitute the "real" voice of the customer (Schmitt 2014). The fashion of explicit and implicit customer requirements is shaped by the individual customer's bygone experience, future expectations, actual needs and perception. The customer's perception of a product particularly relies on the human perceptual senses and their interconnection with the human cognitive system. Physiological and cognitive processes simultaneously designate the overall picture of perceived product quality.

83 % of the human perception is based on vision (Braem 2004). The initial visual perception of a stimulus is carried out by the mere visual system (*early vision*). The evaluation of what has been visually perceived is accompanied and influenced by the human cognitive system (Chen 2003). Therefore, the customer's gaze behavior delivers insights into how customers perceive and evaluate a product. Fixation points, direction and duration of fixation and saccades evidence the pathway of cognitive arousal (Nauth 2012). The pathway of cognitive arousal is predisposed by the product as such, its shape and design, its surface components and the individual customer's experience and expectations. As a result, increased cognitive arousal is associated with providing information about explicit and implicit customer requirements. Screen- and glasses-based eye tracking devices visualize the customer's gaze behavior and allude to stages of increased cognitive arousal (see Fig. 15.4). The pictorial manifestation of eye movement allows for further implications against the background of customer-oriented product development and design.

Whereas the human visual perception is concerned with "scanning" a product's visual properties, the human haptic perception provides information about a product's geometric and material characteristics. The geometric characteristics refer to the

Fig. 15.4 Visualization of customer gaze behavior using eye tracking glasses. The *left* picture shows the fixation points and the direction of fixation for four test persons. The *right* picture depicts the duration of fixation for one test person (own material)

size and shape of a product. The material characteristics comprise a product's texture. The texture of a product can be described according to surface features such as roughness, stickiness, friction and stick-slip. In addition, thermal properties, compliance and weight contribute to the customer's haptic perception and evaluation of a product (Ledermann and Klatzky 2009). The exploration of a product's geometric and material characteristics is based on the physical contact between the customer and the product at hand. The nature of physical contact, be it pleasant or unpleasant, leads to increased cognitive arousal. Descriptive and discriminating studies guide the process of haptic exploration and facilitate the measurement of haptic perception and evaluation while mapping stages of increased cognitive arousal to distinct surface properties (Clark et al. 2008). The focus lies on examining and labeling the customer's hedonic perception during the phase of physical contact.

Electroencephalographic devices (EEG) connect the stages of increased cognitive arousal during visual and haptic perception to participating brain areas (*neurofeedback*). The visual information is transformed into electronic signals leading to enhanced brain activity. The measurement and location of enhanced brain activity provides insights into brain areas participating during the processes of product perception and evaluation (Babic and Damnjanovic 2012). Likewise, devices quantifying the amount of electrodermal activity (EDA) evidence stages of increased cognitive arousal. The human vegetative nervous system controls for perspiration (*biofeedback*). Hence, ascending perspiration accounts for increased cognitive arousal (see Fig. 15.5) (Wagner and Kallus 2014).

Besides sensory information also verbal information is taken into consideration for the purpose of customer-oriented product development and design. The customer's perception and evaluation of a product becomes manifest in textual form appearing as recommendations or complaints. Since social media applications are the primary means of communication nowadays, customers articulate their recommendations or complaints referring to a given product using online communication platforms (Mast 2013). 62 % of all potential customers consult online reviews for advice prior to deciding in favor for or against the purchase of a product (Lightspeed Research 2011). 81 % of all potential customers base their purchase decision solely on the reception of prior customer reviews (E-Tailing Group 2007).

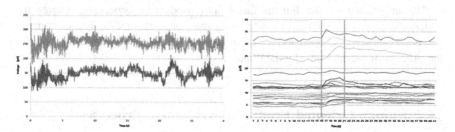

Fig. 15.5 Visualization of electroencephalographic and electrodermal activities. The *left* picture shows the electroencephalographic activities for 2 test persons. The *right* picture depicts the electrodermal activities for 21 test persons during the evaluation of differing sound signals. The *two yellow lines* indicate the occurrence of noise distraction (own material)

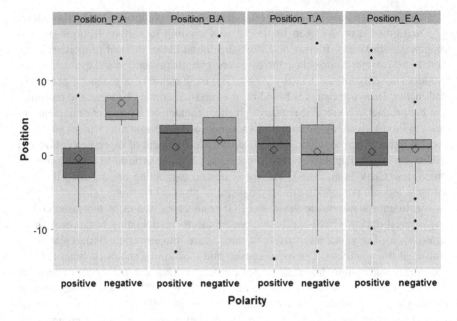

Fig. 15.6 The influence of polarity on attribute position relating to keywords referring to products, brands, types or elements (own material)

The integration of social media applications during the process of customer requirement analysis provides several benefits with respect to the number of available data, the actuality of data and the authenticity of data. The number of available data necessitates the automatic detection, extraction and analysis of relevant user generated content. Opinion mining tools facilitate the above named process relying on machine learning or lexicon-based approaches (Liu and Zhang 2012; Manning et al. 2008; Baccianella et al. 2010). Moreover, the integration of individual grammatical features which constitute the overall linguistic structure of either recommendations or complaints has the potential to increase the accuracy of existing opinion mining tools (see Fig. 15.6).

The integration of the human factor into production processes ensures the realization of products which exactly serve the customers needs. The customer's perceptual senses, cognition and communication behavior deliver valuable input for the process of product development and design.

15.4 Human Oriented Product Development Processes

While the first chapters were addressing the consideration of human factors in the early phases of product realization processes, the company oriented perspective has to consider the human factors of the product development team itself.

The new product development process (NPD) is one of the most important and complex business processes. In order to compete in globalized markets it is necessary to develop products within short time periods and a defined quality level (Barclay et al. 2010). For complex products, the team of a single product development project can exceed easily the scope of a small and medium enterprise by its own. Because of the high amount of functions and people involved (e.g. systems engineering, mechanical engineering, software development, electronic development, project management, product management, industrial engineering) the NPD is characterized by a high complexity, non-linearity and permanent iteration which drive the affordable level of communication and coordination to an extreme. This state is difficult to be controlled (Loch and Kavadis 2008). As illustrated in Fig. 15.6 the methods and activities of quality management are aiming towards a collaborative management of the maturity levels in the fields of project, product, process and contract management.

That is, a new product development project will be successful if the information of the involved actors is allocated in an efficient and effective way, minimizing the amount of failures and iterations due to rework of tasks and assignments. The distribution of information and synchronization of different product and process releases (the maturity levels) is one of the biggest challenges for quality and requirements management. Experts in product development know: the higher the dependence and connectivity of information, the more challenging the planning affords for the information distribution and synchronization between tasks and actors are. Especially delays can cause massive instabilities in the product development system due to an increasing amount of rework and failures (Schmitt and Stiller 2013). The information dependence of requirements and information is illustrated in Fig. 15.7.

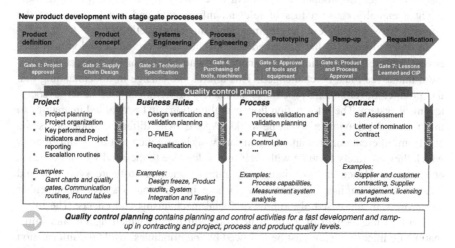

Fig. 15.7 Quality planning and control in new product development processes

If for example the industrial engineering is informed with a certain delay about a change request of the mechanical component, a later change can cause massive extra costs and delays.

Furthermore, the different functions are working according to different development paradigms and models. That is, the project management might stick to the stage-gate-planning, while the systems engineering is applying the V-Model and the software engineering might even develop according to the rules and procedures of agile development methods such as SCRUM. Therefore the communication and synchronization of different functions and actors is even complicated, since the models cannot always be harmonized and synchronized easily. In order to support the activities in product development and realization processes, various expert tools and software systems are used besides informal and formal communication links and channels (e.g. meetings, logs, records, minutes, mails). Requirements management systems, change management software tools and product lifecycle management systems are powerful tools supporting the product development activities within their domain. Within the field of a single domain the applied software systems must be chosen specifically depending on special functions and domain specific requirements. That is the reason for the necessity of using different tools for quality inspection planning (MS-Project, MS-Excel), risk analysis and for production control planning (e.g. CAQ-Systems, SCIO, APIS).

Last, but not least, the employees tend to get lost within the complexity of product development systems and tools. The amount of different methods can hardly be understood and not even be overseen by a single product development team member. Nevertheless the state-of-the-art software systems and methods are by majority emphasizing the system and workflow view instead of putting the employee and his human factors in focus. The allocation of information based on defined rules, regulations is more important than considering the individual competencies and characteristics of the employees. Due to the described complexity of the organization of product development processes and teams, the planning tools might eventually cause a loss of relevant information and can hardly cross the boarders between different development domains and functions.

In order to improve the described state-of-the-art of product development processes, a more human oriented understanding and perception of product development initiatives must be challenged. This affords the change in understanding that the product development process is not solely a technical, but rather a socio-technical system. The information flow and allocation can be regarded in analogy to the rules and behavior in social networks: Agents, with different characters, skills and from different cultures and domains are working on the creation of information which they are likely to share with their principles, due to their individual problems and interest. Hence, a more integrated, human oriented software system would create a social engineering community where information is spread using either workflow and rule based algorithms or social mechanisms and effects. The information-pull mechanism of a social network will decide which agents and information are important and must be followed by the principles, while the information push channels are securing the minimum level of standards and workflows.

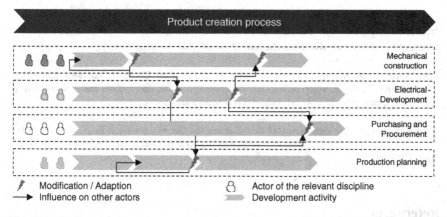

Fig. 15.8 Quality planning and control in new product development processes

Figure 15.8 illustrates the general concept of the social engineering network. When an agent generates information which has importance for a principal within or outside the development team (e.g. line manager, company experts) an algorithm will distribute the information based on a set of parameters (e.g. organization, functions, workflows, dependencies, risks, behavior of employees) (Fig. 15.9).

The design of new product development theories and systems, integrating the existing methods and tools will be one of the great challenges for major improvements in product development processes and for the optimization of the

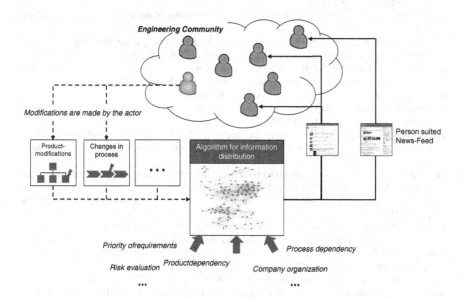

Fig. 15.9 Social engineering network and distribution of information

company perspective in engineer-to-value product management. Moreover, the described methods for the identification and transformation of the customer's perceived quality can increase the value of products significantly.

Acknowledgment The authors would like to thank the German Research Foundation DFG for the kind support within the Cluster of Excellence "Integrative Production Technology for High-Wage Countries.

References

Babic M; Damnjanovic V (2012) Application of neuroscientific concept in defining an effective marketing communication strategy. In: Jakšić M; Barjaktarović Rakočević S (eds) Proceedings of the XIII International Symposium SymOrg 2012. Innovative Management and Business Performance, p 1113–1119

Baccianella S, Esuli A, Sebastiani F (2010) SentiWordNet 3.0. An enhanced lexical resource for sentiment analysis and opinion mining. In: Proceedings of the 7th International Conference on Language Resources and Evaluation,Valletta, p 2200–2204

Barclay I, Dann Z, Holroyd (2010) New Product Development, p 1–3

Braem H (2004) Die Macht der Farben. Müller/Herbing, München

Brecher C, Klocke F, Schmitt R, Schuh R (2014) Integrative Produktion: Industrie 4.0 Aachener Perstpektiven. Shaker Verlag, Aachen, p 200–202

Chen Z (2003) Attentional focus, processing load, and stroop interference. Perception and Psychophysics 65, p 888–900

Clark S, Costelle M, Bodyfelt F (2008) The sensory evaluation of daily products. Springer, New York

E-Tailing Group Inc. (2007) Social shopping study. http://www.internetretailer.com/2007/11/08/customer-reviews-influence-cross-channel-buying-decisions-study. Accessed 21th Aug 2014

Hamouda A, Rohaim M (2009) Sentiment classification of reviews using SentiWordNet. The Online Journal on Computer Science and Information Technology 2(1), p 120–123

Homburg C (2008) Kundenzufriedenheit. Gabler, Wiesbaden

Ledermann S, Klatzky R (2009) Haptic perception. A tutorial. Attention, Perception & Psychophysics 71(7), p 1439–1459

Lightspeed Research (2011) Consumer reviews and research online. http://www.lightspeedresearch.com/press-releases/when-was-the-last-time-you-made-a-purchase-without-researching-online-first/. Accessed 24th Sep 2014

Liu B, Zhang L (2012) A survey of opinion mining and sentiment analysis. In: Aggarwal C, Zhai C (eds) Mining text data. Springer, New York, p 415–463

Loch C, Kavadias S (2008) Handbook of New Product Development Management, Routledge, p 178–182

ISO/IEC: 15288:2008 Systems and software engineering. System life cycle processes

Manning C, Raghavan P, Schütze H (2008) Introduction to information retrieval. CUP, New York

Masing W (2007) Das Unternehmen im Wettbewerb. In: Masing W (ed) Handbuch Qualitätsmanagement, 5th edn. Hanser, München, p 3–14

Mast C (2013) Unternehmenskommunikation aus dem Jahr 2012. UTB, Konstanz

Nauth D (2012) Durch die Augen meines Kunden. Praxishandbuch für Usability Tests mit einem Eyetracking System. Diplomica, Hamburg

Schmitt R, Falk B, Quattelbaum B (2009) Product quality from the customers' perspective. Systematic elicitation and deployment of perceived quality information. In: Huang G, Mak K; Maropoulos P (eds) AISC: proceedings of the 6th CIRP-sponsored international conference on digital enterprise technology. Springer, Berlin, Heidelberg, p 211–222

Schmitt R, Amini P (2013) Analyzing the deviation of product value judgment. In: Abramovici M, Stark R (eds) Smart Product Engineering. Springer, Heidelberg, p 765–776

Schmitt R, Stiller S (2013) Introduction of a Quality Oriented Production Theory for Product Realization Processes, in: Proceedings of the 5th International Conference on Changeable, Agile, Reconfigurable and Virtual Production (CARV 2013), Munich, Germany, October 6th–9th, p 309–314

Schmitt R (2014) Die aktive Gestaltung der wahrgenommenen Produktqualität. In: Schmitt R (ed) Perceived Quality. Subjektive Kundenwahrnehmungen in der Produktentwicklung nutzen. Symposion, Düsseldorf, p 13–35

Schmitt R; Amini P, Falk, B (2014) Quantitative analysis of the consumer perceived value deviation. In: Proceedings of the 24th CIRP Design Conference, Milano, p 14–16

Schuh, G (2012) Innovationsmanagement. Handbuch Produktion und Management 3, 2. Aufl. 2012. Berlin

Wagner V, Kallus K (2014) How to involve psychophysiology in the field of transportation. Recent contributions to an applied psychophysics problem. In: Stanton N, Bucchianico G, Vallicelli A, Landry S (eds) Advances in human aspects of transportation, Taylor & Francis, p 413–417

Printed in the United States
By Bookmasters